The workings of our minds and hearts

Francis Kung   PhD

Revised 17 Nov 2014

Copyright © Francis Kung 2014

All rights reserved. No part of this publication may be reproduced, stored in a retrieval system or transmitted in any form by any means without the prior permission of the copyright owner.

ISBN-13: 978-1500748654

ISBN-10: 150074865X

Printed in the United States of America

Cover design, graphics and layout - Francis Kung

TO MY LOVING WIFE, PHYLLIS

Without you, none of this would have been possible.

ACKNOWLEDGMENTS

I would like to express my appreciation to all who believe that I have something worth listening to. Thanks for your time, kindness and tolerance with errors of any kind you might come across.

Sincere thanks to Karen Moss for her kind offer to read the manuscript. I am so grateful to have a friend like Karen.

Contents

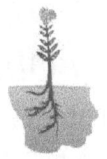

Preface   I
About the author   IX

1 Understanding ourselves   1

Moment of truth   3
Looking inward   10
Being honest with ourselves   15
We are deeper than we think   19

2 The human brain   23

The three brain modules   25
How we learn?   36
Brain plasticity   39
Use it or lose it   47
Who is in charge?   53
What we think we become   59
Multi-tasking   63
How should we use our brains?   66

3 Our minds and hearts 73
The obvious is not easy to understand 75
A belief is just a belief 79
Emotions are just emotions 84
Free association 88
The lie that heals 91
Our memories are malleable 96
Our early years are critical 101
Our stories 105
Why some are unable to tell their stories? 109
Everything that happens leaves a trace 117
Healing through rewiring our brain pathways 125
How do we see the world? 136

4 Learn more and do more 139
We do not know anything 141
We need teachers 145
Mindfulness 150
Say no to past conditioning 156
Whom to marry? 159
A relationship is a spiritual path 162
Control our emotions 168
Free ourselves from bad habits 172
Forgiveness is healing the past with love 175
Let go of our egos 183

5 Think more — 185

- Think differently — 187
- Develop insights — 190
- Understand how the world works — 194
- Nothing will be perfect — 197
- Fleeting beauty of life — 201
- Death, loss and grief — 204

6 Become more — 209

- Open our hearts — 211
- Open our minds — 218
- Hold onto our integrity — 225
- Connect to our calling — 228
- Grow our souls — 231
- Become a bit kinder — 238
- What we need is humility — 243

Author's note — 249

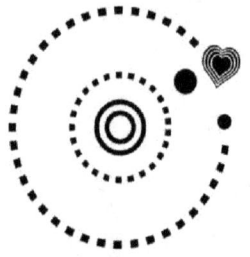

Wisdom is not a product of schooling but of the lifelong attempt to acquire it.

Albert Einstein

Preface

If you have ever wondered why your life isn't unfolding the way you expected or wanted it to, it is because we know very little about ourselves, how and why we think, feel and behave the way we have been.

Most of us know what we should be doing, but we can't always manage it, because we understand very little of how our minds and hearts work. Much of our mental life is beyond conscious awareness because we are influenced by factors of which we are unaware. Our motives are driven by unconscious desires. Therefore, writing this book is both a challenge and equally a worthwhile task.

This book is for those of us who want to live in a more positive and effective way, to make good use of our potentials, and to create a better life for ourselves and others. It is my hope that this book will inspire others to invest time, energy and resources in their own personal development through prioritizing people before things, soul before ego, and inner beauty before external appearances. It is all about how we can tap into our inner goodness and become a bit kinder.

Modern life and scientific progress have changed our style of thinking and perspective. Science helps us to distinguish between myths and inaccurate beliefs from facts, and demystify the mysterious. Brain research findings offer possibilities for us to develop our potentials through better understanding of how our brains work. However, there are limitations of scientific research. Science cannot tell us everything about ourselves, nor give us answers to life's big questions such as what it means to be human or the purpose of life.

> *Knowing others is wisdom, knowing yourself is enlightenment.*   Laozi - Chinese philosopher

In Chapter One – *Understanding ourselves*, we shall explore the journey to understand ourselves. Our realisation usually requires us to look deep inside ourselves, to be honest with ourselves, and allow the universal sufferings and grief of life to invade our minds and hearts. Suffering is indispensable to the quest for wisdom.

Deep down we are mammals with unconscious instincts and drives. In some ways, we are divided against ourselves; different parts of our brains are locked in conflict. The deep unconscious natural processes that are built in by evolution often propel us to act impulsively, selfishly and unkindly. We are selfish because we want to protect ourselves; it is our survival instinct.

However, we are given a layer of rationality which enables us to exercise some restraint over our unconscious instincts and motives. It is our analytical brain (cerebral cortex), the most advanced module of the human brain that can find meaning in life. It can arouse the moral centres and direct us to aim for positive emotions based on being of higher purpose, with the focus on the spirit instead of our ego. When we reflect on the true meaning in life, it can bring out our inner beauty.

In Chapter Two – *The human brain*, the evolutionary and neurobiological models are used to explore the functions, uniqueness, capabilities and potentials of the human brain.

The revolutionary discoveries from recent brain research show that the human brain has the capacity to change its structure and function in response to experiences. This is referred to as neuroplasticity. The human brain can change itself because it is plastic throughout life. This special feature enables us to survive in a changing world, to learn and change so as to create a meaningful life for others and ourselves.

Our brain is influenced by what we do and think. Thoughts hold as much power as actions. Our thoughts serve to lay down the pathways in our brains. Our thoughts can initiate chemical responses in parts of the brain, which in turn activate emotional reactions and subsequently behavioural responses. From the neuroscience perspective, imagining an act and doing it are not as different as they sound because the same parts of the brain are activated. There is scientific support of what Buddha said, "All that we are is the result of what we have thought. The mind is everything. What we think, we become."

Our brain enables us to learn and survive by changing itself. The better we understand our brain's special features and make full use of its potentials, the more rewarding our lives will be.

In modern complex societies, interdependence is a necessity, and our success in life depends to a large extent on our ability to collaborate. Our brain has the potential to help us to achieve higher levels of psychosocial competence. We are not born with wisdom, but our brain enables us to learn and adapt continuously. Insight, generosity and tolerance are not inborn traits, but they can be learned and developed.

In Chapter Three – *Our minds and hearts*, we shall explore the hidden dimensions of our brains - the unconscious mind. It is the powerful force that shapes our minds. Once we have a grasp of our unconscious, it would be like having found a compass that may help us to change and navigate our lives in a positive direction.

Have you ever experienced the feeling that a minor degree of difference is perceived as a world of difference? When we understand why we react to some events as though they are major losses that is the insight that we need to develop, the epiphany of our lives.

Sigmund Freud, the founder of psychoanalysis, developed his ground-breaking theory of the mind in the 1890s. However, Freud's work is often dismissed as being non-scientific, because his investigations were mostly based on single case studies. In addition, Freud's concepts are rather abstract and difficult to measure. If Freud had access to modern brain imaging and brain research technology, he would have proven part of his model on a neurobiological basis.

Actually, as early in the 1960s, there were significant advances in brain research. Therefore, I shall attempt to integrate the findings of brain research with the psychoanalytical model of the unconscious mind, with the aim to gain a better understanding of how our minds work. The ideas presented are sure to change over time as scientists delve deeper into the workings of the mind.

Chapters Four to Six are based on integrative knowledge from the preceding chapters, together with my personal experience and philosophy, which I grouped according to my motto, "Inspire others to think more, learn more, do more and become more."

Transforming our thinking, and subsequently our emotions and feelings, is the most effective way to bring about happiness for ourselves and for all others. Happiness and inner peace comes from being open to our emotions, open to the present moment, open to life, open to our hearts and minds. Spiritual awareness cannot be accessed from the ego.

It is unlikely that someone would come up with the truth that would suit everyone. Hopefully, you will cultivate the seeds of change and take actions to make lasting changes so you can make your life well lived. When you develop your own road map and start your journey, you will find that your life will not be quite the same as before. You have to trust your own experience to guide yourself.

> *We do not need to be a very intelligent person; what we need to be is a kind and loving person; that is a wise person.*

I hope this book is not just paper and ink, and that you will find some of the printed words speak to you, comfort you, open your hearts and minds, and inspire you to build a more meaningful life. I hope you will be inspired to ask yourself, "What can I do to make my life better and that of others?" If you are not ready for change, I hope you will revisit and may we meet again.

> *Books that influence us are those for which we are ready, and which have gone a little farther down our particular path than we have yet got ourselves.* E. M. Forster

Our depths are man-made, which are engraved by kind thoughts and actions. It is through personal and spiritual development that we become deep. We acquire depth through showing compassion and unconditional love in our relationships with others. A small change in our habits in the right direction will benefit others and us in a huge way.

At the end, the testament to a life well lived is the proof that we loved and loved well through the full use of our hearts and minds.

Francis Kung PhD

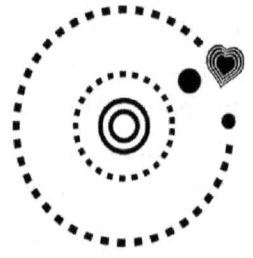

The best teachers teach from the heart, not from the books.

About the author

Dr. Francis Kung's personal mission is to inspire others to think more, learn more, do more and become more.

He does not follow any specific religious ideology. However, he is open to philosophical ideas from different belief systems and is comfortable putting them together in a way that guides his personal growth. Initially trained as an occupational therapist, Dr. Kung then lectured at the Hong Kong Polytechnic. He attended the University of Melbourne for his doctoral research in the area of chronic pain management.

He uses short stories and concludes each article with a take-home message in his books to speak to our hearts and minds. It is his contribution and responsibility to others.

Dr. Kung currently resides in Melbourne, Australia. He devotes time caring for those who have temporarily lost their directions, and sharing insights he has learned from interacting with others.

Wisdom is when we can laugh at our own folly rather than blaming others.

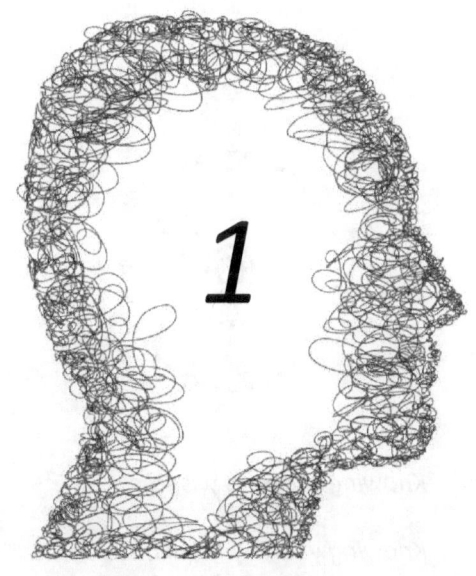

Understanding ourselves

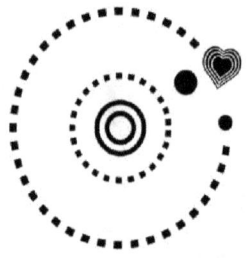

Knowing others is wisdom.

Knowing yourself is enlightenment.

Laozi - Chinese philosopher (sixth century BCE)

Moments of truth

During my travels in October 2010, I joined a tour from Kuala Lumpur to Malacca, Malaysia. I met a fellow traveller from Italy, who worked as a journalist in Asia. At that time, I just finished my first book – *Ways to better living*. As the fellow traveller expressed interest, I gave him a copy of the book. I did not hear from him since then. On 8 March 2014, I received an e-mail from this fellow traveller, in which he expressed gratitude for the book I gave him. He told me in his darkest moments he suddenly remembered the book I gave him and found it to be very helpful. He told me he had not suffered enough in his life to motivate him to read even beyond the first page of my book any earlier.

Nearly all religious traditions put suffering at the top of their agenda. Aristotle believed that tragedy can educate our emotions and teach us to experience them appropriately, and learn to feel compassion for the unfortunate. It is through tragedies that we may then see our own troubles in perspective. Suffering is indispensable to the quest for wisdom.

Show me a hero and I will write you a tragedy.
F. Scott Fitzgerald

The tragic hero in Greek mythology, Oedipus, fulfilled the prophecy that said he would kill his father and marry his mother.

Oedipus was born to King Laius and Queen Jocasta of Thebes, but they abandoned him in the mountainside to die so as to avoid the prophecy. However, Oedipus was found by a shepherd and was raised by King Polybus and Queen Merope of Corinth.

When Oedipus learned about the prophecy, he thought his fate was to kill King Polybus, so he left Corinth. On his way to Thebes, he argued and fought with an older man, which resulted in the older man's death. Oedipus did not know that the old man was King Laius, his father. At that time, Thebes was at the mercy of the Sphinx, and Oedipus was able to defeat the monster, so he won the throne of Thebes and married the king's widow Jocasta, unaware of her true identity as his mother.

Subsequently, Oedipus was determined to search for who had killed the late King Laius, and then discovered he had killed his own father. Oedipus then blinded himself and put himself into exile. Jocasta hanged herself upon realizing that Oedipus was her son and had killed Laius.

The pain of others can become an education in compassion. However, most of us would rather push it away and pretend that the ubiquitous grief around the world has nothing to do with us. Some of us may deliberately steel our hearts against involvement with other people's suffering.

What is the purpose?

The purpose is to create a more satisfying and fulfilling life than we presently experience. It is through doing our best for others, who are stricken by suffering. Our purpose is to live a life filled with passion and meaning.

What causes it to happen?

All I can provide is my own experience. After struggling with my own darkness, I realised the power and hazards of our unprocessed pain. Subsequently, I learned what it takes to make peace with the past. It is about integrating our anger, discontent, fear, insecurities and unresolved conflicts into our conscious mind.

We need to stay in touch with our feelings. Much of our pain and suffering is due to our attempts to dodge difficulties rather than to deal with them head on. We need to face our pain rather than numb ourselves to our true feelings. We need to devote time to work through our anguish and loss. It is also necessary to indulge our pain as fully as possible. We can learn a lot through the unexpected benefits of life's difficulties, setbacks and imperfections.

It was my inability to function well that forced me to strengthen my spiritual connection. All my flaws have brought me priceless gifts, for they are what have led me to reflect deeply on the purpose of life. Ultimately, this led me create a more meaningful purpose beyond what I could ever have imagined for myself, and to use my gifts to benefit others.

How do we do it?

An event only really becomes important to us when we are personally touched by it. When it makes us feel a sense of deep personal concern, it will force us to look at ourselves and also forces us to change.

We should not keep suffering at bay and maintain ourselves in a state of heartlessness. We should open our hearts to the grief of others as though it were our own. If we can allow it to become habitual, this empathic attitude would inspire us to devote our lives to the alleviation of others' hardships.

> *The heart that is never exposed to loss cannot know tenderness.*   Rainer Maria Rilke

We cannot even begin our quest until we allow the universal sufferings and grief of life to invade our minds and hearts. Sometimes, we have to lose ourselves to find ourselves. We need to experience 'the dark night of the soul' in order to start the path to wisdom.

First, we need to detach from the 'me first' mentality.

Second, it is important to revisit our own past pain. The sufferings we have experienced in our own lives can also help us to appreciate the depths of other people's unhappiness. Our own tragedies give us an entirely new vulnerability and consequently, the ability to enter into the suffering of others. It may be our childhood experiences of being neglected, orphaned, parcelled out to relatives, or marginalised from our family.

Third, we need to appreciate the plight of another person. We try to think about the perception of others' perilous situations and feel the impact.

I was asked on a number of occasions where did I learn the skills to understand others. I answered, "I learned from my own suffering. I am one of them." Having said that, I am fortunate to have met a lot of people who have been kind to me.

How it is that someone can suddenly become different?

People do have experiences in which their whole worldview can change dramatically. It is the moment we change our perception and sense of self. We realise that this is the way that it actually is, rather than the way that common sense presents it.

I went on a spiritual journey to find out what I was feeling. The journey required me to give up my work so I could have time to do the work my soul could no longer live without. Actually, that was the best decision I ever made. The search has led me to an unexpected journey and I have found treasures along the way.

Suddenly, I found my passion, which is to do what I can to relieve the suffering of others. It is through developing an open heart we can develop the ability to appreciate the suffering of others.

I had a mystical experience after the occurrence of a very dark night of the soul. The experience brought with it a sudden and powerful experience that occurred literally overnight. It instantaneously and radically changed what I had always known my life to be into something different. I was a different person with a different agenda in life. I awoke with knowledge that I had not previously known. It was the simple, yet powerful concept of *agape* - the love of one's fellow man as one loves oneself. It taught me that we are all one.

Where does this come from?

Recent research suggests that we empathise with other people's pain by simulating their experiences in our own brains. This is referred to as transcendence, the spiritual connection with others, in which we apply the wisdom that we are all one. Transcendence is attaining transpersonal consciousness.

It is through better understanding of ourselves that we acquire wisdom.

We do not need to wait until we are lost in order to start finding ourselves. Remember the things that help us when we are having a bad time – a kind word, a smile and understanding, and try to give that gift to others who are suffering.

In the end, it is through our altruistic actions resulting from an open heart and mind that we can love and inspire others.

When we reflect on the true meaning in life, it can bring out our inner beauty.

1 UNDERSTANDING OURSELVES

Looking inward

It can be self-destructive when we begin to believe things about ourselves that are false. I once watched the Australia's Got Talent (a reality TV show), in which there were contestants who seemed to be quite deluded about their skills and talents. They all had the dream to be a famous singer or performer. However, some had never accurately taken stock of their abilities and talents. Some were unable to accept honest or constructive feedback, and some found it hard to swallow. One contestant, who performed poorly, was asked by the panel judge, "Why do you think you are a marvellous singer?" He replied, "My girlfriend tells me that I am a great singer." The judge then commented, "My girlfriend tells me that I am sexy."

On the other hand, there are those who are amazingly talented but do not realise they are, and perhaps they do not have anyone to encourage them to recognise their strengths.

In the story of *The Wizard of Oz,* the wizard who was just an ordinary person led the characters out of their delusions. The wizard gave the scarecrow, who thought he needed a brain – just a head full of bran, pins and needles. He gave the tin man, who thought he needed a heart - just a silk heart stuffed with sawdust. He gave the lion, who thought he needed courage - just a potion of 'courage.' In fact, each of the characters already had what they wanted within themselves.

When we can fully understand our strengths, we can make the most of them. When we fully understand our weaknesses, we can work on improving them and avoid repeating mistakes. We need to seek honest feedback without getting emotional about the results.

A lot of times, our deepest convictions are based on faulty foundations. Unless we interrogate our most fundamental beliefs, we would live a superficial and expedient life. Socrates wrote, "An unexamined life is not worth living." Therefore, the goal is to change and create a new and more authentic self.

Was your life a success?

The society we live in has its determination of the meaning of good life, but it may not be meaningful. If our sense of who we are is based on simply following others rather than on our true selves then when challenged, we can easily be thrown off track. It is difficult to measure whether we are caring parents, or a good husband or wife. Equally, it is difficult to know whether we choose the right work. A well-lived life can mean different things to different people - success at home, work or leisure. Actually, we live two lives; one is the material existence and the other is the inner life of our souls.

The pace and tone of modern life can make trying to connect with our own inner true self feel like trying to have a deep conversation about a sensitive topic with a 14 year-old who is deeply engaged in a computer game. If we want to understand ourselves, we must begin to cut out outside 'noise' and look inside ourselves.

> *Conversation enriches the understanding; but solitude is the school of genius.*
>
> Ralph Waldo Emerson

There may be a lot of information coming from the outside, but we need to understand the motivations, emotions, impulses and biases that lie within us. This requires us to spend time away from the crowd. This provides the space for us to learn how and why we think, feel and behave the way we have been. Like fishing with a rod and line, we spend time silently sitting and waiting patiently for our unconscious mind to surface. We then need to check with what we think in our head, the logical reasoning mind. There are so many decisions to make, choices to make and relationships to negotiate. Therefore, we need to understand our internal navigation (the unconscious mind), which is in addition to our logical reasoning. This way, our views and decisions will be more balanced.

We converse with others to sharpen our minds, but it is only in solitude that we find ourselves. For those who have a faith, they may converse with the divine in solitude.

Keeping a journal of our thoughts and feelings is a good way to find out our thoughts, passions and personal concerns over a specific time period. Journal writing is a powerful tool; it is intensely helpful. I write to record memories and feelings as an avenue for emotional release. When I read my work again, it helps me describe and reflect on events that have happened. It brings clarity to thoughts and feelings I do not understand. Upon reflection of the written journals, we can figure out what is important. Through the journals, we can look back on our lives, and it can bring order to our experiences and generate awareness and insights.

Think about a time you have cried. It might have been while you were watching a movie, listening to a song, looking at a picture, reading a book or hearing someone say something to you. What was that moment that caused you to be emotional?

We are moved to tears by something at a deeper level when it means a lot to us. If we can figure out why we were moved, it would help us to recognise the things that mean a lot to us, the things we are passionate about and the things we care about. We need to keep asking why it is important to us.

However, those who are ego-centred and short-sighted generally have to experience failure or breakdown on some level before they look at themselves and understand the mistakes they have been making. Sometimes, it is only by doing something wrong that we can learn to do it right. We all make mistakes. Otherwise, we would be like robots, programmed to perform specific tasks without capacity of further development.

> *Your sacred space is where you can find yourself again and again.*   Joseph Campbell

Our purpose is to be discovered within us. In figuring it out, we can discover more about ourselves as well as what matters to us. It is never too late to figure out what is important. It all starts from working with our minds and hearts.

Whatever you are looking for can only be found inside of you.   Rumi (1207-1273)

1 UNDERSTANDING OURSELVES

Being honest with ourselves

We think our children will not tell a lie. Then we find out they told us a lie and we are surprised they had done so. Actually, when the opportunity presents itself, most of us cannot help lying for our self-interests. We will be surprised when we realise what lies beneath us.

Our emotional brain is genetically coded to respond for self-survival. The less than desirable qualities are lying dormant within us and can come forth at anytime, anywhere. That is why at times we become our own worst enemies. We may not be consciously aware of all the qualities we possess. In reality, within every one lies every quality known to humankind. We are kind and warm as well as cold-hearted and selfish.

We all have faults, ourselves, people we feel neutral towards, the ones we like as well as the ones we find objectionable. If we have inherited certain undesirable traits, we need to know. If this is true then we should not reject it simply because it is unpalatable. Rather, we must learn to improve it.

Most of us are blind to the entirety of who we are. The journey to uncover the vast and often hidden darkness that influences our choices can be confronting. The journey will take us into the heart of the duality of darkness and light that operates within each one of us. We need to accept ourselves as we are.

It is not by accident that each of us becomes the person we are today. We have to learn to understand that it is the underlying unconscious mind that pushes us to struggle with the issues that we are faced with. We need to understand it is our unconscious mind that leads us down a dark road to nowhere. We need to unravel the lies and distortions, the guilt and shame that unwittingly turn us into our own worst enemy. The underlying unconscious mind is the culprit that causes us to act out in inappropriate ways, destroy our relationships, sabotage our dreams and place ourselves in harm's way. We need to interrupt the internal mechanisms that cause us to turn our backs on ourselves and give our control away to some thoughts or beliefs.

Generally, we are more afraid of losing what we have than not getting what we want. The willingness to confront what we fear is the touchstone for our personal development. It requires us to step outside what we have long held to be the truth about ourselves and expose the hidden mechanisms that drive us to hurt ourselves and other people.

To be honest with ourselves is one of the greatest risks. It requires courage to visit the places where we have been in denial. When we identify our shortcomings, we bring compassion to the parts of ourselves we have been ashamed of. We also bring courage to the areas of our lives where we have been afraid to admit our vulnerabilities.

> *Whatever that begins with a lie rarely ends well.*
> Gordon Livingston

The less willing we are to change; the less able we are to comprehend the mistakes we have been making, as obvious as they might be. Sometimes, deep personal concern arises when a person has to acknowledge having made a mistake. It is a profoundly unpleasant feeling because it challenges the way we have been thinking, feeling and behaving.

We need to admit to what things we do that are costing us, harming others and ourselves, and how they lead us to becoming our own worst enemy.

When we free ourselves, we shall have the resources to make better choices, think more empowering thoughts, and behave in ways that leave us feeling proud and inspired.

We have to allow ourselves to heal our deepest regrets, challenge our insecurities, befriend our self-doubts, confront our inner demons, and face up to the ways in which we participate in our own self-destruction. We hold the key and are capable of opening up the door to living a life beyond the limitations of our false self.

The process of understanding and learning from the mistakes and bad decisions we made will give us the tools we need to change our lives in the most positive direction imaginable, and also the lives of the others.

We achieve inner peace not by learning new tricks to hide our imperfections but by embracing more of our insecurities, more of our shame, our fear and our vulnerabilities. This is the only path to heal our own suffering.

Conquering oneself is more difficult.

Confucius – Chinese philosopher (551-479 BCE)

Our aim is to see things exactly as they are, without any self-deception or illusion. This will enable us to follow the liberated path.

It is through scrutinizing our path that we may change and grow.

1 UNDERSTANDING OURSELVES

We are deeper than we think

The strictly evolutionary view of human nature sells humanity short. It leaves the impression that we are just slightly higher animals. Human nature that originates from evolutionary biology is not deep, but we can develop ourselves to become a person with depth.

We all are rooted in nature and we originate with certain biological predispositions. However, great men and women whom we admire have surpassed nature. They develop depth by fighting against natural evolutionary predispositions of selfishness.

As a child, we begin with and are influenced by evolutionary forces. If we grew up in cultures that encouraged us to take a loftier view of our possibilities and love of others, we may be able to start the journey earlier. For most of us, our understanding and development of our depth occurs later in life.

Depth is the core of humanity. It is something we cultivate over time. We are not deep when we are young, but we can be as time goes by, depending upon how we have chosen to lead our lives.

How do we invest in our own personal and or spiritual development?

We need to allocate time to reflect, read books, learn from teachers, attend workshops and seminars or perhaps pray and meditate. We need to treat others kindly and give ourselves for the benefit of others. We will continue this journey only if we can wake up to the fact that our personal and spiritual development appreciates day by day, whereas material possessions depreciate.

Our realisation usually requires suffering. Often, it is during moments of suffering that we discover that we are not what we appeared to be. The suffering scours away a floor inside us, exposing a deeper level, and then that floor gets scoured away and another deeper level is revealed. Finally, we get down to the core of human transcendence.

We acquire depth through showing compassion and unconditional love in our relationships with others. We turn the core piece of ourselves into something stable by holding onto our integrity and values even in challenging situations.

It is through making commitments to our nation, faith, calling, loved ones and others that we become deep. We then endure the sacrifices those commitments demand. Those who are deep also made commitments to projects for the benefit of others that extend beyond their lifetimes.

Our depths are man-made, which is engraved by kind thoughts and actions. We carve out depths with an open mind and open heart according to the quality of the commitments we make for the benefit of others. Our depth is also built through suffering, which we are free to choose.

The analytical layer (cerebral cortex) of the human brain provides us with great human possibilities. It is through personal and spiritual development that we become deep. The path of service is very often the path of healing and happiness. We have the potential to make ourselves deep in a way that is beyond the imagination of evolution.

However, the changes we have made to ourselves are only one generation deep. Every generation has to learn anew and afresh.

We all have the potential to become deep, it is our choice how we shall live our lives.

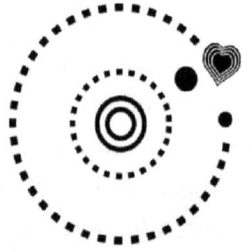

At the end, it is through our altruistic actions resulting from an open heart and mind that we can love and inspire others.

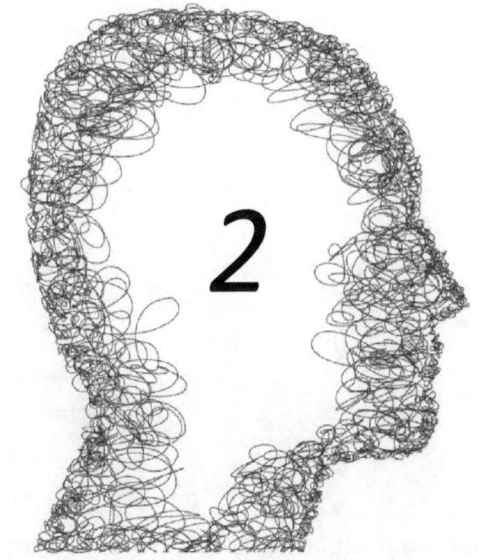

The human brain

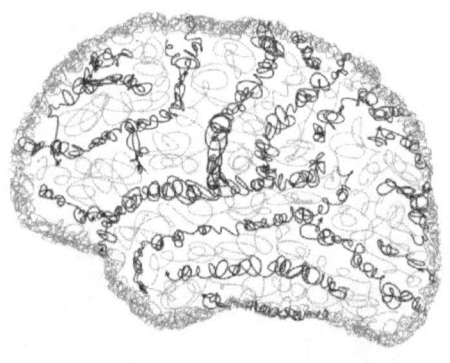

Our brain is plastic. It enables us to learn and adapt continuously so that we can create a meaningful life for others and ourselves.

2 THE HUMAN BRAIN

The three brain modules

The Triune brain model [1] is a helpful schema that can provide us with a simple overview of the human brain and understand its components, functions and the ways they interact. The uniqueness of this model is its simplicity, in which it divides the human brain into three modules based on the evolutionary development of the vertebrate brain. According to this model (Figure 1), the three main modules of the human brain are:

1. brain stem and cerebellum (reptilian brain),
2. limbic system (emotional brain), and
3. cerebral cortex (analytical brain).

An extensive network of nerves connects the three modules of the human brain such that each component influences the other. The three human brain modules operate somewhat independently at times, but they

---

[1] MacLean, P, 1990, *The Triune brain in evolution: Role in Paleocerebral Functions*, Springer, USA.

mostly interact with each other. We can assume that at times one particular module may be dominant while the other modules act in support. In order to function optimally, the different modules need to communicate with each other so as to achieve balance.

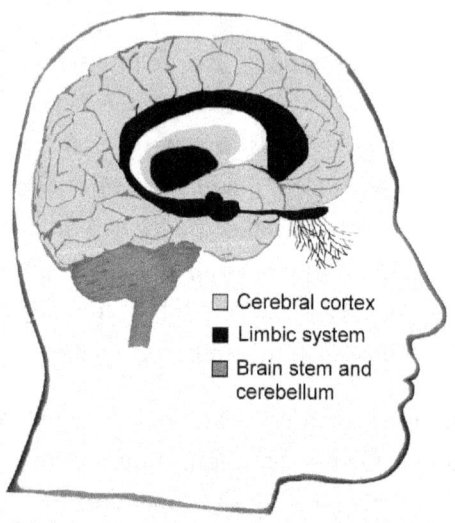

Figure 1 - The Triune brain model

1. The brain stem and cerebellum (reptilian brain)

The reptilian brain is the most primitive part of the human brain. The role of the cerebellum is to coordinate movements. The brain stem deals with vital functions for physical survival and maintenance of the body such as digestion, breathing, circulation, heartbeat and temperature regulation. The brain stem carries out a set program of responses when we are under stress. Any shock or threat will pour adrenaline into our system to prepare for immediate action as required. The 'fight or flight' response is as old as the human race. It is evolved to deal with life back then for our survival from predators. Hence, it is automatic and is highly resistant to change. It is almost outdated in modern societies, though there are still some uses such as when we need to run away from the path of a speeding car.

The reptilian brain responds instinctually and automatically for our survival.

2. The limbic system (emotional brain)

The limbic system is the second part of the brain to evolve. It is basically under the control of the genes, and develops early all by itself. It is referred to as the emotional brain, which deals with unconscious memory, emotions, motivation and emotional responses such as fear. The emotional brain is also believed to be the seat that enables the instinctual protection and nursing of the young.

The limbic system includes a cluster of structures that lie on both sides of the thalamus beneath the cerebral cortex. It serves to enable emotions to be consciously felt by the cerebral cortex and the conscious thoughts to affect emotions.

Generally, emotionally traumatic experiences are more likely to be remembered. Therefore, it is believed that experiences of past traumas are stored in the limbic system. As part of the survival function, the emotional brain associates events in the present time with similar ones in the past, so that the reflexive response will be activated for efficient functioning. For example, if one was attacked whilst walking in the bush one evening, in addition to the instinctual caution, every trip in the bush in the evening will be associated in one's mind with danger. It associates the bush and nightfall with peril. It is a learned reflex, which then becomes automatic such that it becomes almost instinctual.

Our emotional brain instinctually responds by over-estimating threats. Hence, it allows the unconscious memories to be aroused by any emotions, sensations or thoughts that mildly resemble the past experience. For instance, someone with the similar look as our ex partner who betrayed us would immediately draw our disdain. For those soldiers suffering from post-traumatic stress disorder, the sound of a car backfiring may plunge them into thinking they are in the middle of a gunfight. For those with anxiety or panic disorders, their symptoms can be triggered by situations that only remotely resemble the objects of their fear. Our emotional brain is easily scared, so when we perceive a threat, the emotional brain will send out chemicals to trigger the 'fight or flight' response before we have time to determine if the threat is real. Advances in brain scan technology[2] can show the parts of the brain being activated when we are engaged in certain activities, thoughts and moods. For example, hatred activates the amygdala of the limbic system.

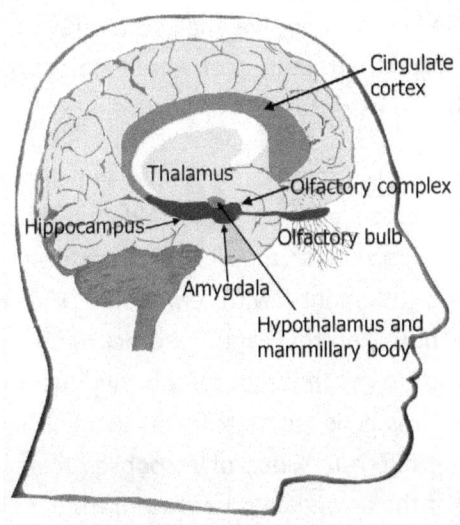

Figure 2 - Components of the limbic system - emotional brain

---

[2] Conventional MRI (magnetic resonance imaging) was first developed in the 1970s and the first clinical MRI of the brain was carried out in early 1980s. The development of functional magnetic resonance imaging (fMRI) in the 1990s provides a tool to gauge real-time brain activity by measuring changes in blood flow.

The limbic system (Figure 2) includes the hippocampus, amygdala, anterior thalamic nuclei, septal nuclei, fornix, mammillary body, olfactory areas, cingulate cortex and parahippocampal cortex. Some experts would also include parts of the thalamus, hypothalamus and midbrain reticular formation as part of the limbic system. In this section, we shall focus the coverage on two important structures of the limbic system, the amygdala and hippocampus.

The amygdala is a small almond-shaped paired structure (mass of nerve cells) on either side of the thalamus within the limbic system, which is responsible for the association of events with emotions. The amygdala assesses both internal and external information for threat and is responsible for alerting us to protect ourselves. It is believed that the amygdala is involved in generating different kinds of responses to fear, and it seems that the amygdala is hardwired to fear for certain stimuli.

The amygdala assesses all stimuli to produce emotional responses. However, we are not aware of it most of the time. For example, when we encounter a frightening sound, it is registered by the amygdala before we are conscious of it. When our analytical brain is assessing the sensory information, the amygdala sends messages to other parts of the brain to trigger changes to prepare the body for the 'flight or fight' response. This is an unconscious reaction and is much faster than the conscious process, which requires the cerebral cortex to recognise the input and extract more information to confirm whether it is a real threat. When the analytical brain realises the sound is harmless, we then relax. This is an example of our conscious response and control of our instinctual response. Not much has changed in the last few thousands of generations as to the ability of our genetic base.

The hippocampus is a horseshoe-shaped paired structure on either side of the thalamus within the limbic system. It is involved in learning, memory and linking memory to emotions. The hippocampus is the gateway through which memories must pass if they are to be preserved. One of the functions of the hippocampus is to turn short-term explicit memory[3] into long-term explicit memories for people, places and things; the memories to which we have conscious access. The hippocampus selects which memories are stored in the conscious level, as we are not able to focus on every piece of information.

---

[3] Explicit memory includes semantic memory that is language-based factual information, and episodic memory that comprises past events, such as experiences, emotions and sensations.

For significant emotional experiences, they are stored as unconscious memories. It is believed that emotional memories may be stored in the amygdala. The hippocampus functions to put emotional markers to emotional memory, e.g. it pulls together the sounds in the auditory cortex and emotions in the amygdala.

Our emotional brain is not based on rationality, it gets hurt and scared easily, and it unconsciously exerts a strong influence on our behaviour based on emotions.

3. The cerebral cortex (analytical brain)

The third and most advanced part of the human brain is the cerebral cortex - the analytical brain. It is the part of the brain that evolves last. The cerebral cortex (cerebrum) is the bulging wrinkled surface we see when looking at the human brain from any angle. It is found uniquely in higher mammals. It is referred to as the grey matter from its colour. Unlike other mammals, humans have a powerful cerebral cortex, where we find the more logical and rational parts of our personality. It is responsible for conscious thoughts and is involved in abstract thoughts, analysing information, reasoning, planning and decision-making.

Most of the human brain expansion is attributed by growth in the connections between brain cells as the result of our experiences, which manifests itself in an expansion in white matter.

There are four lobes in each of the cerebral hemispheres: the frontal lobe, the parietal (middle) lobe, the temporal (side) lobe and the occipital lobe.

At the forefront of the frontal lobe is the prefrontal cortex. The prefrontal cortex is the last part of the analytical brain to mature, and is slow to reach full maturity. Generally, it does not mature until we are in our twenties.

The prefrontal cortex is involved in executive functions such as planning, problem solving, reasoning, acquiring insights, self monitoring, and initiating new and goal-directed behaviour, sustained attention, self-control, flexibility and delayed responding. It also plays a part in encoding and retrieval of memories and is closely linked to emotional regulation. The nerve connections between regions of the prefrontal cortex and the amygdala provide the link to moderate negative emotions.[4]

The analytical brain helps to control and rein in the unconscious and primitive emotions of the limbic system. For example, the analytical brain could consider whether it is appropriate or wise to let out our anger.

---

[4] Banks, SJ Eddy, KT Angstadt, M Nathan, PJ & Phan, KL 2007, 'Amygdala-frontal connectivity during emotion regulation', *Social Cognitive and Affect Neuroscience,* vol. 2(4), pp.303-12.

We use our new analytical brain to step back and become aware of the instinctive and automatic responses of the reptilian brain such as hunger and sex impulses.

An exciting development in brain research since the 1960s is the discovery that the human brain is actually very malleable. Up till the early 1990s, the standard dogma in brain science was that beyond adolescence the brain was set in stone for the rest of our lives. However, recent medical research shows that birth of new brain cells also occurs in adult brains, mainly in the hippocampus,[5] which is central to learning and memory. The human brain is plastic throughout life. This explains why our brain is capable of learning and performing so many different mental functions and behaviour. Our brain's capacity to change its structure and function in response to experiences is referred to as neuroplasticity.

Overall, our analytical brain is mostly un-programmed. The human brain is capable of learning and continuing to be programmed throughout an entire lifetime. This also enables each of our generations to learn anew and

---

[5] Spalding, KL Bergmann, O Alkass, K Bernard, S Salehpour, M Huttner, H & Bostrom, E 2013, 'Dynamics of hippocampal neurogenesis in adult humans', *Cell*, vol. 153(6), pp. 1219-27.

afresh to adapt to the demands of the new environment.

Humans have acquired a complex brain for a complex world, which enables us to survive in a new, changing and demanding environment. For example, on 3 September 1967, Sweden changed the road traffic laws from driving on the left hand side of the road to the right. On that day, all vehicles had to come to a complete stop at 04:50, then changed to the right hand side of the road within 10 minutes and stopped again until at 05:00 before proceeding.

We are not born with wisdom, but our brain enables us to learn continuously and become more. This special feature enables us to learn and change our thoughts and create a meaningful life we want.

Besides the three structural brain modules, our experiences, philosophy and beliefs also interact and affect how we think and behave.

The human analytical brain is a flexible and programmable structure that provides us with the capability for lifelong learning.

2 THE HUMAN BRAIN

How do we learn?

At the physiological and structural level, learning happens when nerve cells connect with each other and form a pathway through which electrical signals flow. Donald Hebb proposed in 1949 that when the nerve fibre of cell A fires and sends an electrical impulse that excites a cell B repeatedly, this leads to increase in synaptic strength between these cells. If the activity is persistent, it tends to induce lasting metabolic changes that add to its stability. Accordingly, nerve structures can be altered by experience, and this provides the biological basis for associative learning. This is now referred to as Hebb's Law, which is often summarised as *'Nerve cells that fire together, wire together.'*

Hebb's Law has been applied to Ivan Pavlov's classical conditioning experiments, which showed that after a bell was paired with food, dogs would salivate upon hearing the bell even when no food was present. This supports that learning changes the connections between nerve cells.

Eric Kandel, recipient of the 2000 Nobel Prize in Physiology or Medicine for his research on the physiological basis of memory storage in neurons, pushed the frontier even further. He proposed that our thinking, learning and behaviour can turn our genes on or off, thus shaping our brain anatomy.[6]

We learn by doing new activities, having different experiences and thinking differently, which then form new connections between different nerve cells. When we repeat certain actions, it helps strengthen existing nerve connections.

Our brains are designed to choose the nerve pathway that requires less energy. The existing pathways are like grooves that make it easier to travel. These are nerve pathways in our brains that are like helpful shortcuts that make life easier. However, some of these nerve pathways may limit us and cloud our outlook.

The brain stores all external information that comes as a pattern in our nerve pathways. This includes not only what we have experienced through the senses externally, but also the sensations, emotions, thoughts and feelings associated with it. When we are challenged

---

[6] Kandel, E 2000, 'The molecular biology of memory storage: A dialogue between genes and synapses', *Nobel Lecture*, 8 Dec, Stockholm.

to learn something new or change our view, this requires more energy, effort and attention.

We can also unlearn by weakening previous learning. When we find that something we believe is no longer the case, we then form a new perspective. The electrical signals of our new thinking will run through a different nerve pathway. When the existing nerve pathway does not receive reinforcement, the connections are weakened.

When we use our brain in a certain way, permanent changes are made to its structure. What we focus on grows. If our daily news is filled with threats, our brains tell us to pay attention and check if there are dangers. The threat alert will keep us watching over our shoulders.

Through our development, we learn the culture of the society we live in. We learn the civilisation, we learn to acquire virtues and inhibit our aggressive and selfish instincts. This will then serve as the reference point for how we will respond in similar situations, often unconsciously. The brain does this so that it can operate more efficiently.

If we train our minds, our brain structures will change based on how we use it.

Brain plasticity

Brain plasticity or neuroplasticity refers to the brain's ability to change and reorganise its functions at all ages and even following damage. Our brains are capable of changing their structure and function in response to our experiences.

In many ways, Paul Bach-y-Rita (1934-2006) was considered the father of the idea of brain plasticity. He was convinced of the plastic properties of the brain long before it was possible to prove they existed. He introduced the idea of sensory substitution as a model and evidence of brain plasticity.

According to Bach-y-Rita, we see with our brains, not our eyes. Our eyes merely sense changes in light energy; it is our brains that perceive and hence see. How a sensation enters the brain is not important.

In the late 1960s, Bach-y-Rita used skin and its touch receptors as substitute for the retina to enable people who had been blind from birth to see. Blind subjects were placed behind a camera, and the camera's image was translated to electrical signals that stimulated 400 small touch sensors that vibrated against the blind subject's back. The pattern of vibration from the 'tactile-vision device' allowed the subject to detect faces, shadows and objects.[7] It was the first application of neuroplasticity, the brain's ability to change. Later in his career, Bach-y-Rita also worked on the application of vestibular substitution for people who are unable to maintain balance.[8]

Our brain is able to use the part that is devoted to processing touch, to adapt to the new signal and decode the skin sensations and turn them into pictures. However, Bach-y-Rita's finding was ignored because it was contrary to the prevalent belief at that time, which was based on the localisation model. The localisation model postulated that each point on the body surface has a nerve that passes signals directly to a specific point on the brain, anatomically hardwired at birth. Each mental function is always processed in the same location in the brain.

---

[7] Bach-y-Rita, P 1969, 'Vision substitution by tactile image projection', *Nature,* vol. 221, pp. 963-4.

[8] Bach-y-Rita, P 2005, 'Late human brain plasticity: vestibular substitution with a tongue BrainPort human-machine interface', Plasticidad y Restauracion Neurologica, vol. 4, no. 1-2, pp.31-4.

With advances in technology, scientists use brain imaging to watch the brain working on various tasks. The purpose is to locate which parts of the brain are primarily related with certain functions. It also attempts to find which parts of the brain give us certain abilities. Brain mapping can also examine how our environment changes our brain's structure by studying how the brain changes physically through the learning processes.

About three decades later, scientists used brain scans to test Bach-y-Rita's concept of neuroplasticity and confirmed that the tactile images could indeed be processed in the visual cortex.[9] Actually, the visual, auditory and sensory cortices are able to process whatever electrical signals sent to them.[10]

Most studies used animal models to explore the concept of neuroplasticity, because research ethics committees are unlikely to approve the experiment in humans.

---

[9] Burton, H, 2003, 'Visual cortex activity in early and late blind people', *Journal of Neuroscience*, vol. 23(10), pp.4005-11.

[10] Finney, EM, Fine, I & Dobins, KR 2001, 'Visual stimuli activate auditory cortex in the deaf', *Journal of Natural Neuroscience*, vol. 4(12), pp.1171-3.

David Hubel and Torsten Wiesel [11] discovered there was a critical period from the third to the eighth week of life, when a newborn kitten's brain had to receive visual stimulation in order to develop normally. They conducted an experiment in which one eyelid of a kitten was sewn during its critical period of development. After a while when they opened the shut eye, they found that the visual areas in the brain that normally processed input from the shut eye had failed to develop. The kitten is blind not in the sewn-up eye itself, but in the part of the brain that is no longer capable of responding to electrical signals from that eye.

However, the brain area that was deprived of visual stimulation did not remain idle. The processing capacity of that part of the brain had been reallocated during the critical four to six-week period of development. Instead, it processed visual input from the open eye. The observation supported the idea that the brain is plastic in the critical period, and it found a way to rewire itself. For this discovery, they received the Nobel Prize in Physiology or Medicine in 1981.

---

[11] Hubel, DH & Wiesel, TN 1970, 'The period of susceptibility to the physiological effects of unilateral eye closure in kittens', *Journal of Physiology*, vol. 206(2), pp.419-36.

Michael Merzenich, one of the pioneers in the field of neuroplasticity, conducted a number of studies in the 1980s that supported the idea the brain can rewire itself. In one study, the median nerve of a monkey's hand was cut. As expected, the brain area (cortical representational zone) for the median nerve was silent when the middle of the hand was stroked. When the parts of the hand served by other nerves (radial and ulnar) were stroked over the study period, the results showed that the brain areas that represented the radial and ulnar nerves progressively expanded to occupy larger and larger portions of the cut median nerve brain area.[12]

Merzenich conducted another experiment that involved amputating a monkey's middle finger.[13] After few months, it was found that the brain area for the amputated finger had disappeared, and the brain areas

---

[12] Merzenich, MM Kassas, JH Wall, JT Sur, M Nelson, RJ & Felleman, DJ 1983, 'Progression of change following median nerve section in the cortical representation of the hand in areas 3b and 1 in adult owl and squirrel monkeys', *Neuroscience*, vol. 10:3, pp. 639-65.

[13] Merzenich, MM Nelson, RJ Stryker, MP Cynader, MS, Schoppmann, A & Zook, JM 1984, 'Somatosensory cortical map changes following digit amputation in adult monkeys', *Journal of Comparative Neurology,* vol. 224, pp.591-605.

for the adjacent fingers had grown into the space that had originally mapped for the middle finger.

In a later experiment,[14] two of a monkey's fingers were sewn together so that they both moved as one. After a few months allowing the monkey to use the fingers, the brain scan showed that the two brain areas of the originally separate fingers had merged into a single brain area. All movements and sensations in those fingers always occurred simultaneously, and they formed the same brain area.

The mainstream belief up to the early 1990s underestimated what the human brain could do. Therefore, traditional rehabilitation usually ended after a few weeks when a patient stopped improving or reached a plateau. However, the period with no further improvement is only temporary. Actually, it is just a stage in the learning process, which is followed by periods of consolidation. Internally, biological changes are happening, as new skills became more automatic and refined.

---

[14] Allard, T Clark, SA Jenkins, WM & Merzenich, MM 1991, 'Reorganisation of somatosensory area 3b representations in adult owl monkeys after digital syndactyly', *Journal of Neurophysiology*, vol. 66, pp. 1048-58.

Paul Bach-y-Rita had firsthand evidence through his father's stroke in 1958 that late recovery could occur even for people with a massive brain lesion. Paul and his brother succeeded in helping their father to lead a relatively normal life despite the opinion of several doctors that it was impossible. When Paul's father died in 1965, the autopsy revealed that he suffered severe damage to a large portion of his brain, which had not repaired itself. However, the fact the he had made a significant recovery suggested that his brain had reorganised itself. [15] This finding has outdated previous belief that there were minimal changes within the brain after adulthood. Review of animal and human neuroscience research shows that a damaged brain can reorganise itself, as reflected in structural changes in the brain. [16]

The belief that treatment for many brain injuries was ineffective or even unwarranted had taken hold for a very long time. However, as technology advances and as we gain a better understanding of the human brain, we

---

[15] Abrams, M & Winters, D 2003, 'Can you see with your tongue?' Discover Magazine, 01 June.

[16] Taub, E Uswatte, G & Mark VW 2014, 'The functional significance of cortical reorganization and the parallel development of CI therapy', *Front Human Neuroscience*, vol. 8, pp.396

have scientific proof that the human brain can change, even in adult life. For instance, treatment based on the outdated belief did not provide the right conditions such as adequate time, motivation and instructions for stroke patients to reclaim maximum lost functions possible. The compensatory approach that relies on the unaffected body part to compensate for the lost function for neurological injuries such as stroke is not the best option. Whereas, the restorative approach based on the principles of neuroplasticity is the better option, as it aims to actively encourage the nervous system to work hard by using the affected arms and legs, thereby building new pathways in the brain.

Because the human brain could change, then people with injury to the brain might be able to form new brain pathways. The application of the principles of neuroplasticity can help people with injuries to the brain form new nerve connections by getting their healthy nerve cells to fire together and wire together.

Many of our brain nerve networks are not hardwired. The human brain is plastic; it is able to change.

Our brain enables us to survive in a changing world by changing itself, even in old age and following damage.

2 THE HUMAN BRAIN

Use it or lose it

When our brain is stimulated, it forms new pathways. A study found that London taxi drivers who must undergo extensive training have an enlarged hippocampus compared to those who do not drive taxis.[17] The hippocampus is the area of the brain deep within the temporal lobe of the brain that is associated with the formation and consolidation of memories and supports spatial navigation. The London taxi drivers are required to use this part of their brain extensively in their daily work. The finding supports the idea that neurogenesis (development of new nerve cells) may be stimulated by learning one's way in a new environment. Learning can stimulate the hippocampus to grow, and this supports the link between exercising a brain function and growth of the region supporting that function.

---

[17] Maguire, EA Gadian, DG Johnsrude, IS, Good, CD Ashburner, J Frackowiak, RS & Frith, CD 2000, 'Navigation-related structural changes in the hippocampi of taxi drivers.' *Proceedings of the National Academy of Science*, vol. 97, pp.4398–403.

Brain scans showed that speaking a second language builds more connections between nerve cells. In addition, there appears to be a relation between grey matter (nerve cell body) and proficiency in a second language. A study found bilingual adults have denser grey matter in the inferior frontal cortex of the brain. This is most pronounced in people who learned a second language before the age of five and were proficient in a second language.[18] A brain scan study also showed that areas activated in bilinguals when switching languages is higher than monolingual individuals.[19]

The way the human brain forms areas to represent the body and allocation for different function is not fixed. Rather, it is based on the principle of 'use it or lose it.' There is competition for using the brain areas. If we stop exercising our mental skills, the brain area for those skills is turned over to the skills we practise instead. If we do not practise our second or third language for a long while, then the brain area is assigned to another skill.

---

[18] Mechelli, A Crinion, JT Noppenev, U O'Doherty, J Ashburner, J Frackowiak, RS & Price, CJ 2004, 'Neurolinguistics: structural plasticity in the bilingual brain', *Nature,* vol.14, 431, pp.757.

[19] Hernandez, AE Dapretto, M & Mazziotta, J 2001, 'Language switching and language representation in Spanish-English bilinguals: an fMRI study', *NeuroImage,* vol. 14, pp.510-20.

The more we use our native language, the more it comes to dominate our brain linguistic area. That is why it is difficult for adults to learn a foreign language. However, for children, if they learn two languages at the same time, during the critical development period, both languages will get a foothold. Brain scans showed that in a bilingual child all the sounds of the two languages share a single large brain area. [20]

Throughout our lifetime, the nerve cells of our brains remain capable of adapting their complex patterns of interconnections to a new condition of use. However, our brains cannot keep on developing if there are no new tasks to accomplish and no new challenges to meet. It is like the saying the rich get richer and the poor get poorer.

The human brain is not hardwired for a lot of activities for modern society such as using a computer for work and leisure. Modern society requires our brains to adapt to new exposures such as reading and using computers.

However, neuroplasticity (the brain's ability to change) can give rise to both flexibility and rigidity. Anything that involves unvaried repetition can lead to rigidity. For

---

[20] Doidge, N 2007, *The Brain that changes itself*, New York, Viking, p.60.

example, if we stay in the same job, doing the same tasks for decades, using the same skills, and attending the same cultural activities, it is not learning. Rather, it is just replaying skills we have mastered or repeating habits. Then, it becomes increasingly difficult for us to change, even if we want to. If we only seek out like-minded people to associate with, we would ignore or discredit information that does not match our beliefs because it is distressing to think in unfamiliar ways.

In extreme situations, some would even try to impose their views on others, such as cults, fundamentalism of religious groups or brainwashing in totalitarian regimes.

From the neuroscience perspective, people can be indoctrinated to the extent that there are anatomical changes in the brain, which makes it very difficult to overcome the differences of opinion with ordinary persuasion.

In a television interview, a teenage North Korean defector said that she and her mother could not believe that Kim Jong-il died in 2011, because he was not a human. He was a God. It took the teenager three years to get over the brainwashing to see the truth.[21]

---

[21] *Insight: Changing a mindset.* 2014, television program, Special Broadcasting Service One, Australia, 8 April.

For some, immigration can be disorienting and traumatising. When we change culture we are shocked because we have to learn to behave differently, something not natural at all. Culture shock demands an intense workout for the brain because it requires massive rewiring. It is not just about simply learning new things because the new culture is competing with the existing nerve networks that were deeply wired and established during the critical period in the native land from which one originates. A similar experience would be like moving to a new house; we have to make the conscious effort to go to a different location to find the bathroom.

When we acquire new experiences, culture and skills, it involves growth; it is referred to as additive plasticity. However, plasticity can also be subtractive, which is taking things away, as in pruning away nerve connections that are not being used. Each time we learn something new and use it repeatedly, some nerve pathways may be weakened in the process as plasticity is competitive. Because of this characteristic of brain plasticity, it is wiser to get it right early before the bad habit gets a competitive advantage. This is why unlearning is often more difficult.

Post-mortem examinations have shown that education increases the number of branches among nerve cells in the human brain. An enriched brain is one that is active, has lots to do, and is challenged, but not overwhelmed.

The human brain can be modified throughout our lives. However, we often underestimate our potential for flexibility.

We have options to take different paths. However, if we keep taking the same path again and again, tracks will start to develop and soon we will tend to get stuck in a rut that becomes self-sustaining.

The way we use our brains, in terms of type and intensity, determines how many connections are built up among the billions of nerve cells in it. The nerve pathways, patterns of connection, complexity and stability also determine how well we have developed our brains.

Doing new, challenging activities can help to stimulate our minds and nurture our growth.

Whoever uses his opportunities will grow.

2 THE HUMAN BRAIN

Who is in charge?

The evolutionary model and the findings of brain research provide the foundation for us to better understand ourselves. Besides the reptilian brain that deals with vital functions for physical survival and maintenance; we basically have two systems inside - the conscious and unconscious.

Deep down we are mammals. Deep in the core of our being there are unconscious instincts and drives. These deep, unconscious, natural processes are built in by evolution in our limbic system (emotional brain), and they often propel us to act impulsively.

Our learned reflex is based on instincts and emotions. It is not based on logic. This can cause us trouble, hold us back, and is the source of our inability to make lasting change. It can cause us to behave in ways that lead to repeated mistakes.

However, unlike other mammals, humans have a layer of rationality in the analytical brain, which serves to support our conscious, rational processes.

The analytical brain functions at its optimal level when it is not stressed. In times of stress, it can clear away the stress, it can choose between shades of grey, it can find meaning in life, and arouse the moral centres with higher meaning. This higher layer does its best to exercise some restraints over our unconscious instincts and motives.

When the analytical brain is dominant over the emotional brain, it can direct us to aim for positive emotions based on being of a higher purpose. The feeling of joy is produced when the conscious thoughts of the analytical brain and the moral and reward centres of the emotional brain are activated. Doing something meaningful or just the thought of being good awakens the pleasure centres of the brain. The focus is on the spirit instead of our ego. The joy based on higher purpose is different from that of hedonistic pleasures, which often leads to overindulgence and ultimately trouble.

The emotional brain is a powerful hidden force that controls us. It makes decisions for us that we don't realise. Decision-making and judgment are profoundly affected by emotions. Emotion drives action; emotion is the power that drives us. Unless we use our analytical brain to communicate and exert control over the emotional brain, we are not able to change, despite how much we want to change. Our emotions may seem to be conscious, but they are actually physiological responses to stimuli designed for our survival – to run away from danger, protect ourselves and act in a way for reward to satisfy ourselves.

If the analytical brain does not take the dominant role, then it does little more than justify afterwards what our unconscious emotional brain has already decided we were going to do.

The source attribution for our unconscious emotional memory is not as obvious as our conscious memory. Therefore, we are not aware that our behaviour is an over-reaction due to past experiences. We do not know what triggered it. For instance, when we first meet a person, we immediately know whether we can trust him or not, though we may not have a rational reason for it.

The human brain is more highly attuned to threat rather than reward. This is the nature of the survival mechanism. The habit of noticing threat more than reward means we often give more energy to the very things we wish to avoid, rather than cultivating the pleasant things that we would like to grow. That is why most of us would rather give up dreams that require us to take a chance on something that bears some risks.

When a stimulus arrives in the brain, nerve pathways compete for arousal; the most dominant nerve pathway wins and fires. That firing strengthens the wiring, and the pathway that loses becomes weaker.

We need the analytical brain to help us to examine the situation and overcome the automatic response and calm our brain when the threat is not real. However, if the emotional brain is dominant, the emotional regulation role of the analytical brain is compromised.

Stress can come from our basic nature, environment, relationships and the way we think and respond. When we are under greater pressure than we can cope with, we shall experience negative symptoms. Stress is cumulative and it follows the pattern of the positive feedback loop, in which there is no shut-off valve. When our stress threshold is exceeded, we lose the resilience to stress. Each episode after the first does not just trigger the stress, but it amplifies all similar experiences.

When we are overly stressed, our thinking is rigid; we do not feel joy. Our thinking and behaviour is reactive, our mood is unbalanced, we are disconnected from others, and our responses can be irrational. We become overwhelmed. We lose our direction and control. Then, even small setbacks and difficulties become major challenges. Eventually, we become worn out.

Stress shuts down our ability to feel and think rationally. It also blocks effective communication with others. The stress response spontaneously induces unkind thoughts about others. We are aggressive because we are afraid, we want to protect ourselves; it is our survival instinct. If we are dominated by anger, we can be destructive. When we get a clear idea of the cause of our emotional distress, then we can take steps to cope with it.

Our analytical brain is like the boat that sits on the surface of the ocean, which can be affected by our emotional brain. If the sea is rough, like when we are stressed, the small boat may not be able to focus, plan and do the fishing. We need a calm sea in order to hook the fish. When we are calm, we are more creative, we can connect to the abstract, we feel joyous and we can open our hearts to connect with others and behave in a kind manner. Our analytical brain has the power to rewire our emotional nerve pathways and oversee the emotional brain, appraise our emotional state, and do what is needed to process the stresses to seek balance.

The two-way traffic between the emotional brain and the analytical brain allows emotions to be consciously felt and the conscious thoughts to affect emotions.

There is some truth in the saying, "As we grow older, we become wiser." Ageing does often result in gaining wisdom. It is because the lifetime experiences can broaden our perspective. We can see the big picture through appreciating the interrelationships and complexities of things.

From the neuroscience perspective, as we age or mature we can pay more attention to bigger picture strategies because of higher levels of understanding, rather than relying on emotions. This is achieved through better inhibition of emotional impulses when our prefrontal cortex acquires dominance over the amygdala (the 'fear centre' in the brain), which alerts us to protect ourselves.

Also, as we mature, there is better communication between the right and left hemispheres rather than dominance in one hemisphere. This symmetry provides a broader perspective, so we can make a better sense of what happened and the demands required of us.

When we are stressed, the more primitive module of the brain is dominant.

When we are balanced and relaxed, the analytical brain is dominant.

What we think we become

Our brains monitor our thoughts, beliefs, emotions and feelings. Our brains also monitor which mindsets are helping or hindering us.

Thoughts hold as much power as actions. If we have angry thoughts, though we suppress them, we are still laying down the nerve pathways that will make it difficult to change.

Our thoughts serve to lay down the nerve pathways in our brain. Our thoughts can initiate chemical responses in other parts of the brain, which in turn activate emotional reactions and subsequently behavioural responses. For example, when our thoughts invoke stress associated physiological and emotional reactions, they can divert us from our usual course of behaviour.

We cannot stop having thoughts; the challenge is learning to have thoughts that have a positive impact on others and ourselves. Our thoughts may be holding us back from achieving our goals. For instance, we tell ourselves that we are too old to learn ballroom dancing or not smart enough to work with computer graphics.

In 1904 Santiago Ramón-y-Cajal[22] proposed that thoughts repeated in mental practice would strengthen the existing nerve connections and create new ones. However, at that time he did not have the technology and tools to prove it.

Neuroscientist Alvaro Pascual-Leone[23] designed an experiment to test whether mental practice and imagination could lead to physical changes in the brain. He taught two groups of people, who had never played piano, a sequence of notes. The first mental practice group just sat in front of the piano, imagined using their fingers to play the sequence for two hours a day for five days. The second physical practice group actually played the music for the same duration. The participants of both groups had their brains mapped before the experiment, each day during the practice and after the experiment. Upon completion, the participants were asked to play the sequence and their accuracy was measured. The results show that both groups learned to

---

[22] Santiago Ramon-y-Cajal was awarded the Nobel Prize in Physiology or Medicine in 1906, together with Camillo Golgi, in recognition of their work on the structure of the nervous system.

[23] Pascual-Leone, A Nquyet, D Cohen, LG Brasil-Neto, JP Cammarota, A & Hallett, M 1995, 'Modulation of muscle responses evoked by transcranial magnetic stimulation during the acquisition of new motor skills', *Journal of Neurophysiology*, vol. 74(3), pp.1037-45.

play the sequence, and there were similar brain map changes. In addition, the mental practice group participants were as accurate as the actual group participants. When the mental practice group finished the training and was given a single two-hour physical practice, its overall performance improved to the level of the physical practice group's performance at day five.

Eric Kandel, recipient of the 2000 Nobel Prize in Physiology or Medicine investigated the physiological basis of memory storage in nerve cells. His investigation shows that as we learn, our individual nerve cells alter their structure and strengthen the synaptic connections between them. Kandel's work[24] shows when we form long-term memories, nerve cells change their anatomical shape and increase the number of synaptic connections they have to other nerve cells. He proposed that when we learn, we change which genes in our nerve cells are turned on so that a new protein is made that alters the structure and function of the cell. We can shape our genes, which in turn shape our brain's microscopic anatomy. The implication is that our brain is influenced by what we do and think.

---

[24] Kandel, E 2000, 'The molecular biology of memory storage: A dialogue between genes and synapses', *Nobel Lecture*, 8 Dec, Stockholm.

From the neuroscience perspective, imagining an act and doing it are not as different as they sound because the same parts of the brain are activated. The same pathways are engaged. Thinking is almost the same as doing. Each thought alters the physical state of our nerve synapses at a microscopic level. Our thoughts leave material traces in our brains. Similarly, everything we do leave traces in our brains.

What fires together wires together. What fires apart wires apart. Try this exercise - write down three people you are furious with. Then, express thanks to the same people who upset you for giving you the opportunity to practise tolerance. If we can show gratitude to the people who taught us tolerance, we can weaken the nerve pathway that causes distress when thinking about the same people.

> *Watch your thoughts; they become words.*
> *Watch your words; they become actions.*
> *Watch your actions; they become habits.*
> *Watch your habits; they become character.*
> *Watch your character; it becomes your destiny.*
>
> Frank Outlaw

Our brain is influenced by what we think and do.

Multi-tasking

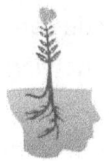

The fast pace of life can be exhausting. However, we might not be able to see that this can cause physical, emotional and mental health problems as we try harder to go faster. We may believe that we should be able to go at a faster pace and there is something wrong with us if we can't keep up. It becomes a vicious circle when we cannot see that we are contributing to our stress levels. The never ending chase and over-scheduling then become chronic stressors that affect our health.

Over-stimulation from mass incoming information through technological changes also contributes to erosion of our attention and creativity. We have less time to reflect on anything, as the urgent need to respond takes over. The flood with massive useless information and pseudo-truths also make it difficult to distinguish the important and unimportant, and the truth from falsehood. With less time for reflective thinking, it is difficult to respond wisely.

There are those who keep rushing frantically until they lose the ability to sit still to reflect or think. Some of us may become excited with trivial matters, and may develop a need to be online and robotically checking constantly. They may surf the Internet for hours without knowing what they are looking for. They log onto chat groups and exchange trivialities with people they have never met. They do not have anything meaningful to say and hence are not able to form close bonds with other people in the real world. Eventually, their perception is blunted through the over-stimulation of forever changing images, noises and sensational inputs. All these demands actually interrupt and diminish learning, productivity, reflective thinking and social skills.

Research studies using functional magnetic resonance imaging show humans don't do lots of things simultaneously.[25] Instead, we quickly switch our attention from task to task. For simple tasks, it may not require high speed switching. For more complex and similar tasks, there is competition for the same nerve pathways and processes. Hence, there will be interference. In most countries, there are road laws that prohibit drivers from talking on their mobile phones, reading or sending text messages whilst driving. These tasks are competing for mental resources, and would

---

[25] Charron, S & Koechlin, E 2010, 'Divided representation of concurrent goals in the human frontal lobes', *Science*, vol. 328(5976), pp 360-3.

over-stretch our brain's capacity to focus. Multi-tasking in this situation can cause serious consequences. Even the use of hands-free mobile phone devices whilst driving can cause deterioration in driving performance.[26]

However, for the students who are proud of their abilities to multi-task; that is listening to music, responding to messages from social chat, emails, phone messages every few minutes, doing their university or school work, and searching the Internet for the best shopping deals at the same time; it is difficult for them to understand that the quality of their academic work will suffer.

We think we are paying attention to everything around us at the same time. But we are actually not. Multi-tasking is a delusion. We have over-estimated our capabilities. We cannot focus on more than one thing at a time. Multi-tasking eventually leads to fragmented attention, which reduces our efficiency and performance.

Multi-tasking is fractured attention. We have over-estimated our capacity to multi-tasking.

---

[26] Just, MA Keller, TA & Cynkar, J 2008, 'A decrease in brain activation associated with driving when listening to someone speak', *Brain Research*, vol. 1205: pp.70-80.

2 THE HUMAN BRAIN

How should we use our brains?

Our brains can be used in ways that depends on how it is structured. On the other hand, how our brain is structured in turn, depends on what it has been used for.

In modern societies, it appears the main focus for most people is on economic success. However, our daily challenges are more of a social and relationship nature. The human brain thrives on enriching relationships.

The most important decisions we have to make in the course of our lives are actually psychosocial in nature, such as our relationships with our partner, family members and others at work and in the community.

At the family level, the system of social relations is getting more complex in modern societies due to changes in family structures and dynamics. We may be part of a blended family or an adopted family. The shift from extended to nuclear families also results in intergenerational relationship changes. In addition, there are new forms of unions, decreases in marriage, and increases in separations.

In the workplace and community, we have to deal with complex demands and relationships. Therefore, our success is more related to our adjustment to the psychosocial demands.

At the macro level, it is our ability to live harmoniously and cooperatively in the community that contributes to the success of human race. We cannot live in isolation.

> *If you want to go fast, go alone. If you want to go far, go together.*   African saying

As life goes on, demands are more complex. Even children are suffering from the effects of our fast-paced, competitive lives. Teenagers have to face the turmoil of transition from being dependent to independence while at the same time coping with educational demands, relationships and career choices. The prefrontal cortex in teenagers is not yet fully developed; hence they are relatively more overwhelmed by emotions and may not be able to form thoughtful responses. Adults face juggling work and family, usually also caring for elderly, frail or sick parents. In late adulthood, we have to face burying friends and family members.

The human brain allows us to adapt to different demands placed on us at different stages of life. Our brain has the capacity to enable us to learn from, understand and interact with our environment. The nerve networks inside our brains adapt to our changing environment, so as to develop the skills to function well in the community. Michael Merzenich, one of the pioneers in the field of neuroplasticity said, "Our brains are different from all those humans before us. Our brains are modified on a substantial scale, physically and functionally. Massive changes are associated with our modern cultural specialisations."[27]

If we do not enhance our capacity to process the increasing complexity and demands of daily life, we shall lag behind and become unable to function effectively and efficiently. It is like an old computer not being updated with the optimal operating system for the new, more demanding applications and programs. Over time, we shall be not able to cope. For instance, analysing facts, data and figures, and using modern technologies such as mobile phones and computers are only recent uses of the human brain.

In order to acquire adaptive psychosocial behaviour for interdependence, we need enhanced communication skills. The human brain has the unique predisposition to generate speech and learn language, which is based on the relationship of symbols and meanings. This enables us to have complex communication with each other,

---

[27] Olsen, S 2005, 'Newsmaker: Are we getting smarter or dumber?' *CNet News,* 21 Sept.

which supports better understanding, collaboration and social cohesion. The capacity to learn and use language separates us from other animals. Thirty thousand years ago humans drew on cave walls. Then, hieroglyphics were invented, which later were converted into letters or characters. Reading involves strengthening nerve connections between different functions that process the images of letters, their sounds and their meanings.

The question we should put forward is - what purposes should we use our brains so the potentialities built into it can be fully actualised?

The purpose of our brain is to help us to achieve a higher level of psychosocial adaptation so we can respond adequately to demands and changes in the environment to ensure our survival.

It is through repetitive use of the new adaptive responses that our brain can lay down new nerve pathways and new behaviours that enable us to achieve higher levels of psychosocial competence.

First, we need to increase our awareness of what influences our emotions and reactions.

Second, we need to develop skills of impulse control, conflict resolution and relationship-building to cope with life's challenges. When we can regulate our emotions, it helps to reduce stress and anxiety.

When we develop better emotional regulation after exposure to conflict and in the face of adversity, we can remain calm and do what it takes to maintain relationships and our inner peace. This is what contributes to our emotional intelligence.

Third, we need to have as many different experiences as possible in our lives with other people who are different from ourselves. This will enable us to acquire broad and comprehensive knowledge, and a wide variety of abilities and skills. When we are no longer dependent on one particular way of living, we can then decide freely how and for what purposes we shall live. If we surround ourselves with people who are wired similarly, then it is not possible to experience new learning.

Fourth, our brains will help reduce the gap between our potential higher nature and our human selves through social bonding, which contributes to the cohesion and sense of togetherness of our communities. At a higher level, we can contribute to the network of interdependence required for advancement of human race.

We can create new nerve pathways that enable our high-level consciousness to bring out our moral consciousness and altruism. Our thoughts can have an effect on our behaviour, and back onto our consciousness. Unless it is strongly wired, under difficult situations our integrity may not hold as evident in situations of theft and destruction during civil unrest or natural disaster.

Insight, generosity and tolerance are not inborn traits, but they can be learned and developed. We can become more. We are not born with wisdom, but our brains enable us to learn and adapt continuously.

We have to teach each generation about our values, to restrain or channel our animal instincts into acceptable expression because what we have developed for ourselves is only one generation deep.

Exercising our minds to its full potential is hard work. However, our analytical brains enable us to gain more control over our automatic reactions. We are deeper than we think.

Our brain enables us to create a meaningful life for ourselves and others. We can be wise and kind, depending upon how we choose to lead our lives.

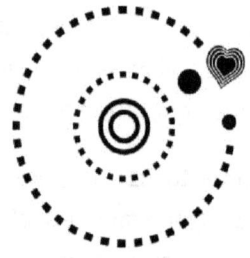

The miracle is not to walk on water or fire.
The miracle is to walk on the earth.

Linji – Zen Master

3

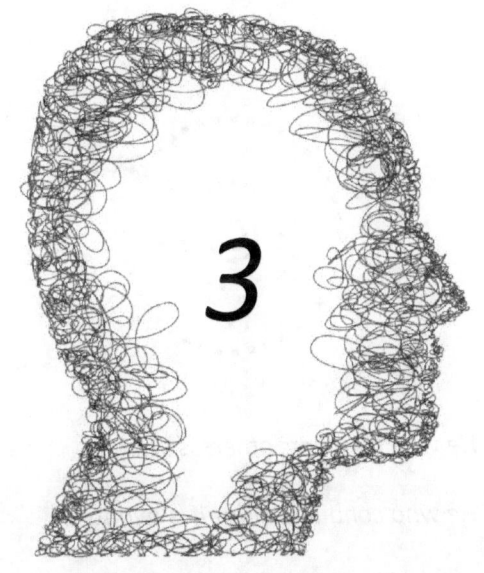

Our minds and hearts

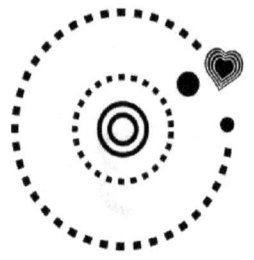

He who conquers others is strong.

He who conquers himself is mighty.

Laozi - Chinese philosopher (sixth century BCE)

3 OUR MINDS AND HEARTS

The obvious is not easy to understand

In Hans Christian Andersen's story – *The Emperor's New Clothes*, the emperor hired two weavers who promised him the finest suit of clothes that was invisible to those who were stupid or incompetent. When the swindlers delivered the invisible clothes, the Emperor and his ministers all pretended that they could see the invisible clothes. The Emperor then paraded before his subjects in his new clothes. Among the crowd, a child cried out, "But the Emperor isn't wearing anything at all!"

I have observed similar situations in the modern workplace. In a mental health facility, the medical team was quite arrogant about their clinical knowledge, skills and judgment. The medical team would not listen to other members of the treatment team, whom they regarded as inferior. The nursing team was in close contact with the patients around the clock. Therefore, the nurses actually had a better idea about the patients' mental states and responses to the medications prescribed. When the nurses reported to the medical team that certain patients were not responding to the prescribed medications or suggested adjustments to the dosage were required, the medical team would not consider their views at all. Numerous repeated attempts by the nursing team to convince the medical team to review the prescriptions failed. It was like talking to a brick wall.

There were situations in which particular patients were at risk due to not receiving the required medications. Therefore, the nurses considered they had obligations to make adjustments based on their clinical judgment and years of experience that were above and beyond that of the medical team. That was done in good faith, though it could be argued that it might have gone beyond the legal and ethical boundaries. After receiving the adjusted treatments, the patients' mental states improved dramatically as expected. Unknowingly, the medical team took the credit and attributed the improvements to their superior clinical expertise. The medical team was unable to learn from others and continued to apply the ineffective treatments all along. Sadly, the senior doctors and consultants also passed the ineffective clinical knowledge onto the junior doctors. The cycle kept repeating itself. The medical team regarded themselves to be great team players. They thought they could work happily with anyone. In reality, they could not work well with the persons with whom they were actually working. The medical team was never ready to receive open and honest feedback about the effectiveness of their clinical skills.

> *The jewel that we find, we stoop and taketh, because we see it; but what we do not see, we tread upon, and never think of it.* Shakespeare

We can apply the same principle to analyse losses in relationships. There are those who simply could not get over a breakup. They may claim no one else could live up to their ex-beloved, and life is not meaningful without their ex-beloved. In reality, we all have the ability to balance ourselves after a certain period of grieving.

Breakups generally involve a degree of rejection. A psychological study[28] suggests that those who overreacted to rejection were more likely to have an extended period of grieving when their partners left them. It is possible that some of those who suffer greatly from a breakup could possibly be attributed to the mindset they hold; that is they could not accept the thought that someone actually left them. The study postulated that these individuals have high rejection sensitivity. The study also found that if the breakup was

---

[28] Ayduk, Ö & Gyurak, A 2008, 'Applying the cognitive-affective processing systems approach to conceptualizing rejection sensitivity', *Social and Personality Psychology Compass*, vol. 2(5), pp.2016-33.

initiated by themselves or by both parties, their grieving will be much shorter.

We are more interested in how others make us feel and what we get from them. It is a problem we create for ourselves. The reason why we find it hard to recognise that our attachment is problematic is because we have a strong tendency to look for faults outside of ourselves. We tend to turn outward to assign responsibility for our problems, and blame our unhappiness on external factors. Given all the pain and confusion our attachments cause us, it is still difficult for us to see that attachment itself is a problem. The problem is not our objects of attachment that fail to make us happy. Rather, it is the feelings of attachment rooted within us.

Often, if we are unhappy we look to see what needs to be fixed or what needs to be blamed. Often, we make changes in our external environment without examining our internal lives, which may have been contributing to the situation in which we were not finding satisfaction. The result is we shall find the same complaints and disappointments in any new environment, which leads to further distress. We must always say what we see, but above all and more difficult, we must always see what we see. As the saying goes, "The fish is the last to discover the ocean."

In finding the truth, we find ourselves.

3 OUR MINDS AND HEARTS

A belief is just a belief

When our minds tell us a story, we believe it. Our minds are the master storytellers. The Chinese ancient fable *Snake in a goblet* [29] is a simple illustration of the concept of imaginary fear.

A court official was invited to the minister's home for dinner. At the table, the court official saw a tiny snake in the goblet, which his superior gave him to toast. He did not dare to refuse and he drank the wine. Upon arriving home, the court official was sick as he felt the snake inside his body. He consulted his doctor and also took medication for his illness. As he was unable to work for a few days, his superior came to visit him and asked about his illness. The court official told his superior that he felt ill because he drank the wine with a tiny snake in his goblet the other day. His superior went home and sat at

---

[29] 杯弓蛇影 - Chinese saying based on the story about mistaking the reflection of a bow for a snake.

his dinner table thinking deeply about the situation. At nightfall, the superior noticed that the bow hanging on the wall of the dining room threw a shadow upon the dining table. He immediately sent his servant to fetch the court official. He sat the court official at the same position at the dining table and offered him a goblet of wine. The court official saw the tiny snake again and was frightened. His superior asked the court official to remain calm and pointed out that the snake in the goblet was nothing but a reflection of the bow on the wall. Straightaway, the court official felt at ease and all his pain and ailments disappeared.

The story shows us that our imagination is sufficient to activate the fear centre of our brain.

We suffer more from imagination than from reality.
Seneca the Elder

If we perceive that the world is dangerous and people are not to be trusted, it will affect how we perceive strangers that we meet. However, if we adopt a belief based on trust, then our perceptions and responses to strangers may be different.

We can conjure disaster from the smallest threat. We often do not realise that a belief is just a belief. However, we often take it as reality and believe that is the way things have to be. This is because we consider our perception as the true perception. We knowingly participate in this one-sided ego validation. It is very powerful and real to us, but it may not be true. Our truths may not be the accurate snapshot of what is going on. They just represent our preferred stories. For instance, there are people who hold an excessive preoccupation with defects in their physical appearance, which is just a slight physical anomaly. When the concern becomes excessive, it can cause distress or impairment in social functioning and quality of life.

Our brains are constantly trying to make sense of the world. One way of doing this is by creating explanatory stories or ideas that fit our experiences. The beliefs we adopt are often useful, but may not always be correct. We need to come face-to-face with our limiting beliefs installed earlier in our lives, often without us even knowing about it.

Our biology, history, experience, upbringing, knowledge and interpretation contribute to the large spectrum of beliefs we adopt. Less organised beliefs usually have a lower status and are often referred to as superstitions, whilst the mainstream beliefs hold a higher status.

If we believe in the supernatural divine, we will 'see' events of coincidences as evidence of our beliefs. Our beliefs will validate the patterns in our minds, and may lead us to 'see' things that are not there; or to make links between events that are not actually connected.

The world is full of uncertainties and unknowns. We can choose to learn to tolerate the uncertainties and the unknown. Some people do not need answers to everything; they are willing to accept the unknowns. Perhaps, the middle road is not a bad choice when we cannot find the truth.

Most of our perceptions are erroneous, and those who suffer most are people who have wrong perceptions. We need to learn to know which perceptions are causing suffering. We need to learn to look with clarity and calmness so as to improve the way we perceive. Once the true reason becomes conscious, it will lose its grip on us. We can then set ourselves free.

Modern science is based on metaphysics of materialism – the idea that the fundamental nature of reality is pure matter or energy. Then, consciousness has no place whatsoever. However, it makes no sense that we get consciousness (non-physical) out of a bunch of ingredients that were entirely non-conscious (physical) to begin with. That would require a miracle, which modern science denies.

There are rational explanation of things and events based on our senses and logic. There are also other ways of knowing, such as feeling and intuition, as well as spirituality. We have to acknowledge that psychic phenomenon is part of the spectrum of human experiences. This is because life is full of uncertainties and unknowns. Science without religion is lame; religion without science is blind.

There is great diversity in the mental inclinations and dispositions of human beings. So there will be a vast spectrum of philosophical, religious and spiritual traditions; it is a valuable heritage of mankind.

The questions we can ask ourselves are:

- What tendencies or experiences in our past have led to our limiting beliefs?

- Are we ready to adopt an open mind to uncover and reprogram them?

We spend our whole lives bending and shaping everything to fit our beliefs. However, we can make choices and we do have free will.

If our brains make up stories and we claim those stories to be the truth, we can also consciously create new stories to rewrite our truths. The human brain is plastic, which enables us to function better through reprogramming our beliefs. It is our choice.

A lot of our beliefs may not be true.

3 OUR MINDS AND HEARTS

Emotions are only emotions

When someone does something significant to us, it will trigger an emotion. When our brain perceives a change in the external world that leads to upset in our internal balance, a feeling arises. This feeling tells us something out there in the world around us or within us has changed.

Emotions are reflected by how we feel. Emotions can be positive such as pride, or it can be negative such as regret, shame or embarrassment. Unfortunately, sadness, anger and fear are the emotions that get triggered most often. Emotions can amplify our experiences because they are also associated with physiological changes and thoughts. This may affect how we respond to the event that triggers our emotions. We are all unique; that is why one situation affects one person positively, but might affect another negatively.

Some emotions can be disorientating. They can divert our attention from objective logical thinking and important issues. Then, we are not able to see the big picture. Some emotions can freeze us or impel us towards irrational behaviour. When our emotional brain is dominant, we cannot think clearly about cause and effect. Therefore, it is not wise to make important decisions when we are in our worst moods. When we are angry, we may say unkind words or act in a manner that hurts others and causes damage to our relationships. All of which eventually will hurt ourselves.

On the other hand, others can use emotions to exploit us. For instance, the salesman preys on our fear of losing a bargain. This opens the opportunity for the salesman to keep using the same old tactic, 'Today only special,' to push us to finalise the purchase as soon as possible. It is similar to the common emotional trick that online dating scammers use, claiming their close family members are very sick and need money for urgent medical treatments.

If we are unfortunate to have parents or close family members with sociopathic traits, they can easily play emotional tricks on us. When a manipulative mother plays the emotional trick, and says to her adult child, "I am your mother. I will not do anything to harm you. Don't you trust me?" Most of us would fall into the trap, even if we consider ourselves to be fairly intelligent. This is partly because we are conditioned to love our parents. We are brought up to respect our parents, and we are brought up to have absolute trust that they will not harm us. However, this may not be true. Since we have been brainwashed or conditioned through our culture, acting against one's culture, upbringing, moral and social values is a very difficult task. It requires wisdom to see through the emotional tricks others use upon us, especially those who are close to us.

Another emotion we have to deal with is fear. Fear is a natural reaction to our vulnerability to loss. What are we afraid of? Actually, our biggest fear is nonexistence after death, and it happens that major religions use this as the foundation by promising us some versions of immortality after our extinction.

Fear springs from ignorance. The less information we have about something, the more threatened we are likely to feel about it. When we understand more, we shall fear less.

There are other types of fear as part of living in modern society, such as fear of humiliation and fear of failure. Sometimes, we are not able to fulfil our dreams because of our fear. We may have to ask ourselves, where is the energy tied up in our lives? Is it spent running away from our fear?

> *The worst sorrows in life are not in its losses and misfortunes but in its fears.*   A.C. Benson

Sometimes we do not have a specific object to fear. We just keep worrying, showing physiological symptoms of anxiety. We will do anything to avoid situations that trigger anxiety. This is the more efficient way for the brain to minimise anxiety. However, this method that requires less effort is only a short-term relief. Actually, attempts to evade our fears make them worse. The only way to overcome fear is to confront it. The common psychological approach for treatment of phobia is based on the classical conditioning theory that uses a hierarchical approach of applying gradually more intensive and repeated exposure to an anxiety-provoking stimulus (using real or guided imagery) in a safe environment, paired with relaxation training and deconstructing the evoking process.

My son had an adverse reaction to flying. Even the thought of flying would trigger adverse reactions, both emotional and physiological. Over the years, instead of avoid travelling by air, he insisted to travel with us, initially on short flights and gradually on longer flights. It was a long process and finally he was able to travel on his own and be comfortable flying without fear.

Emotions affect our body, thinking and behaviour. We cannot stop having emotions; the challenge is learning to understand our emotions, regulate the negative emotions and act wisely.

Free association

Sigmund Freud (1856-1939), the founding father of psychoanalysis was involved in physiology research for some years before specializing in neurology and setting up his private practice to treat psychological disorders. In the early years of his laboratory research, Freud proposed the concept - *nerve cells that fire together wire together*,[30] which was expanded subsequently and is now referred to as Hebb's Law.

Freud also proposed the concept of *synapse* – which is the small gap separating nerve cells that contains neurotransmitters at the pre-synaptic ending and receptor site at the post-synaptic ending. That was a few decades before the advent of Sir Charles Sherrington, who was regarded as the first person to introduce the word 'synapse' in 1897. Freud also put forward the idea that the synapse might be changed when we learn.

---

[30] Sacks, O 1998, 'The Other Road: Freud as neurologist', in M Roth (ed.) *Freud: Conflict and Culture,* Knopf, A. New York, pp. 221-234.

Donald Hebb[31] developed a theory of how nerve connections form and how synaptic connections change with learning. Hebb proposed that when the synapse of one neuron repeatedly fires and excites another neuron, there is a permanent structural change that strengthens the synaptic connection. Hence, repeated stimulation of a group of nerve cells can lead to the formation of networks of nerve cells that persist even after stimulation has stopped. The more times synapses in these assemblies fire together, the stronger the network becomes, which increases the likelihood they will fire together again.

These scientific findings may serve to explain Freud's technique of 'free association,' in which patients in psychoanalysis lie on the couch and 'free-associate.' Patients are encouraged to say everything that comes into their minds, regardless of how uncomfortable or trivial it seems. The most important thing is that the logical mind does not intervene to censor spontaneous thoughts. The loose approach of letting patients go in any direction may generate random thoughts. A more commonly used technique is the controlled approach, in which a broad topic such as family is given to the

---

[31] Hebb, D 1949, *The Organization of Behavior*, New York, Wiley & Sons.

patient. Another technique used is when the therapist mentions a single word, and the patient spontaneously responds with the first word that comes to mind. For example, the therapist said, 'mother' and the patient immediately responded with 'witch'.

Freud used dialogue with patients as the clinical method for treating psychopathology. Through the dialogue Freud found interesting connections emerged in his patients' associations; the thoughts and feelings the patient normally pushed away into the unconscious. In our daily language, we often humorously refer this as a Freudian slip, a verbal mistake that reveals some type of hidden emotion on the part of the speaker. For example, in 2012 a news channel presenter in the United States introduced Prince William as 'the douche' of Cambridge and he then hastily corrected it as Duke of Cambridge. According to Freud, such errors may reveal an unconscious thought, belief or wish.

From the neuroscience perspective, the nerve cells that fired together years ago are wired together, and these original connections are often still in place and can show up in a patient's free association. Every thought we have leaves a trace.

Our mental associations are expressions of links formed in our memory networks, they may not be as random as they seem.

3 OUR MINDS AND HEARTS

The lie that heals

Our mind and body are inextricably linked. The mind body link can be well illustrated through the 'placebo effect'. Usually the term 'placebo effect' refers to the helpful effects a placebo has in relieving symptoms. However, the placebo effect usually lasts only a short time.

A placebo looks like an active drug, but has no pharmacological properties of its own. Yet, mysteriously, when given under the right conditions or by the right doctor or healer – a placebo can affect healing.

Placebos seem to affect how people feel. About one out of three persons experienced a change in symptoms as a result of getting a placebo. Because placebos often have an effect, even if the effect does not last long, some people think the placebos produced a cure. However, placebos do not cure because they do not act on the disease.

Sometimes if the placebo looks more real, the person may think it is an active medicine or treatment and believe in its power even more. For example, a larger pill may look more powerful than a small pill. And in some people, an injection may have a stronger placebo effect than a pill.

If the patient believes in the treatment and wants it to work, it can seem to do so, at least for a while. Some believe placebos seem to work because many illnesses improve over time even without treatment. Hence, some scientists believe the effects of many alternative therapies may simply be a placebo effect as well.

In 1957, Mr Wright was in the final stage of lymphatic cancer and only had a few weeks to live. Mr Wright was desperate and he begged and pestered his treating doctor, Dr Philip West to give him a new drug he heard would be the miracle cure. However, the new drug was only offered for clinical trials to people who have at least three months to live, for which he was not qualified. Eventually, Dr West gave in and reluctantly administered the new drug to Mr Wright. To Dr West's amazement, Mr Wright was without symptoms within ten days and he was discharged home.

Two months later, Mr Wright learned from scientific literature that the miracle drug did not seem to be effective. Mr Wright was depressed and had a relapse.

Dr West genuinely wanted to help Mr Wright, so he told Mr Wright the scientific literature did not have complete information about the drug. Dr West then advised Mr Wright that he had a new batch of the refined drug, which was double the strength. Dr West then gave Mr Wright the new refined drug, which was actually an injection of distilled water. Two months later, Mr Wright had a dramatic recovery.

Dr West insisted that Mr Wright needed to continue the treatment for some time. During that period, Mr Wright remained symptom-free. However, subsequently Mr Wright read a conclusive report from the American Medical Association that the drug was totally ineffective in the treatment of the type of cancer he had. Mr Wright's condition then deteriorated and he died two days later.

In 2013, Ambulance Victoria (Australia) discovered that someone had tampered with vials of the potent painkillers (morphine or fentanyl) and replaced the contents with saline in one of the service regions.[32] In the preceding year, two paramedics from Ambulance Victoria were arrested after hundreds of vials of fentanyl (a highly addictive opioid pain killer) went missing and

---

[32] 'Victorian ambulance patients given saline instead of painkillers', *Australian Broadcasting Corporation news online*, 20 Feb 2014.

were replaced with tap water.[33] As other paramedics were not aware of the drug switch, they unwittingly administered water to patients in the mistaken belief they were providing the pain relief required. However, before the discovery of the drug switch, no complaints were received from ambulance patients claiming inadequate pain relief.

For psychological disorders, particularly depression,[34] it has been shown that placebos can be nearly as effective as active medications.

A double-blind, placebo-controlled study that used a PET (positron emission tomography) scan to observe metabolic changes in the brain showed that patients with major depression who responded to a placebo and those who responded to an anti-depressant had similar metabolic changes in the cortical and limbic regions of

---

[33] Bucci, N & Gordon, J 2012, 'Paramedic questioned over theft of ambulance painkiller', *The Age newspaper*, 17 Oct.

[34] Wampold, BE Minami, T Tierney, SC Baskin, TW & Bhati, KS 2005, 'The placebo is powerful: estimating placebo effects in medicine and psychotherapy from randomized clinical trials', *Journal of Clinical Psychology*. Jul, vol. 61(7), pp.835-54.

the brain.[35] Based on the findings, the researchers commented, "Treatment with placebo is not absence of treatment, just absence of active medication."

It is generally agreed placebos cannot cure a disease or illness. However, this is an important area of study as we can learn more about how placebos may be helpful for different conditions through better understanding of the neurobiological mechanisms.[36]

> *Our minds are bound to our bodies like an oyster is to its shell.* Plato

What we think does make a difference to what happens in our brains and body.

---

[35] Mayberg, H Silva, A Brannan, S Tekell, J Mahurin, R McGinnis, S & Jerabek, PA 2002, 'The functional anatomy of the placebo effect', *American Journal of Psychiatry*, vol. 159, pp.728-37.

[36] Benedetti, F Mayberg, H Wager, T Stohler, C & Zubieta, J 2005, 'Neurobiological mechanisms of the placebo effect', *Journal of Neuroscience*, vol. 25(45), pp.10390-402.

3 OUR MINDS AND HEARTS

Our memories are malleable

We are unable to remember any of the experiences we had before the age of two, because the prefrontal cortex would not have matured enough to effectively store the experience in the conscious memories.

Sigmund Freud believed that events we experienced can leave memory traces in our minds, which can be altered by subsequent events. For instance, young children who were molested might not be able to understand what was being done to them, but once they matured, they might look upon the incident anew and give it new meaning, and the memory of the molestation changes.

In 1896, Freud proposed that memory traces are subjected to rearrangement in accordance with fresh circumstances. In order to change the memory, Freud argued that the memories had to be conscious and become the focus of our conscious attention. Unfortunately, some of the traumatic memories might have happened very early in childhood, which are not easily accessible to consciousness, so they are very difficult to change.

Recent research shows that the details of childhood memories cannot be accurately recalled. Therefore, our brains fill in the gaps with fabricated details. In reality, our memories of past events are re-created each time we remember them.

Our memories may not be true to the actual event because they are malleable, and can change as new information comes in.

Generally, we think a detailed account of events based on one's memory is probably true. However, this may not be the case. This is because we will not be able to remember all the details. Therefore, some parts of the contents are filled with reconstructed memory rather than facts. Even accounts given by confident individuals with emotional attachments may not be the total truth of what had happened.

A study that used suggestive interviews to ask participants to talk about events that happened in their childhood showed that some participants were led to believe that they had been lost in a shopping mall and rescued by an elderly person.[37] This finding suggests our

---

[37] Loftus, EF & Pickrell, JE 1995, 'The formation of false memories', *Psychiatric Annals*, vol. 25, pp.720-5.

memories can be susceptible to distortion and perhaps manipulation by others.

The implication is that it is plausible that stories can be planted into others and some may claim horrific memories of abuse were repressed until their therapies. In the courtroom, it is equally possible that a well-meaning prosecutor and or a co-witness may produce information that gets blended with another witness' memory of what happened. Generally, our faulty memories are not going to cause major issues. However, in the courtroom it can cause travesties of human justice. Therefore, some courts of justice have set guidelines on how to evaluate eyewitness evidence. Psychologists are called upon as experts to provide juries with information on how memory really works, because not everyone knows how memory works.

This reminds me of David Pelzer's autobiography, *A Child Called It*, which is a memoir of child abuse.[38] The book described in exceptional details about what had happened over a long period of time. David's younger brother, Richard had also published a book, *A Brother's*

---

[38] Pelzer, D 1995, *A Child called It: One child's courage to survive*, Health Communications, USA.

Journey,[39] which wholeheartedly supported David Pelzer's version of events. However, David's three older brothers felt David had exaggerated the abuse, despite strong evidence supporting what had happened. This illustrates that our memories could be affected by our perceptions or perhaps be subjected to manipulation by others.

Greater understanding of the memory processes has prompted scientists to look for a method to erase unwanted, unpleasant memories, so as to prevent post-traumatic stress disorder.

Scientists have found propranolol (a beta blocker used for heart conditions) binds to the same cell-surface receptors as adrenaline and noradrenaline, which are released when the amygdala is activated by threatening situations.

An early study[40] found the group of patients who took propranolol after traumatic events had a far lower rate

---

[39] Pelzer, R 2006, *A brother's journey: surviving a childhood of abuse*, Grand Central Publishing, USA.

[40] Vaiva, G Ducrocq, F Jezequel, K Averland, B Lestavel, P Brunet, A & Marmar, CR 2003, 'Immediate treatment with propranolol decreases posttraumatic stress disorder two months after trauma', *Biological Psychiatry*, vol. 54(9), pp. 947-9.

of showing symptoms of post-traumatic stress disorder compared to the group who refused.

Other studies using propranolol in conjunction with psychotherapy also showed promising results. In those studies, subjects participated in script-driven mental imagery of their past traumatic events to help reduce the impact of traumatic memories formed years ago.[41]

There are also studies that used different medications for the treatment of post-traumatic stress disorder. The common feature is these chemicals can affect the processing and creation of long-lasting fearful memories stored in the hippocampus.

The focus of future studies regarding post-traumatic stress disorder is on prevention as a realistic goal.

We can consciously re-create new stories to rewrite our truths.

---

[41] Brunet, A Orr, SP Tremblay, J Robertson, K Nader, K & Pitman, RK 2008, 'Effect of post-retrieval propranolol on psychopshysiologic responding during traumatic imagery in post-traumatic stress disorder', *Journal of Psychiatric Research,* vol. 42(6), pp.503-6.

Our early years are critical

Everyone is aware that there is a critical period for efficient learning, for example the learning of languages. Early exposure to two or more languages results in children becoming multilingual as easily and naturally as they learn their 'native' language. However, for an adult to learn another foreign language, the task is much more difficult and rarely as well achieved.

Similarly, there is a critical period in our psychological development. According to Freud's psychoanalytical model, the critical period for the development of our ability to love is in early childhood. What happens during the critical period has an inordinate effect on our ability to love and relate later in life. It is still possible to make changes later in life, but it is much harder after the critical period ends.

The earliest connection between parent and child sets the foundation for establishing the nerve pathways into the unconscious memory system of the emotional brain of the child. From 10 to 18 months, the right frontal lobe of the infant is developing. This part of the brain has a significant role in social and emotional behaviour, in terms of maintaining human attachments and regulating emotions. During this critical period for emotional development and attachment, parents will be teaching the child what emotions are by using speech and nonverbal communications. The child will learn the name of the emotions, what causes it, how one feels, the associated bodily sensations and how to relate to others. Through this, the child will learn to know and regulate his or her emotions and be socially connected.

Social deprivation at certain critical periods in early childhood could have negative effects on children's subsequent development. Rene Spitz's studies in the 1940s compared infants reared by their own mothers and those in a group home.[42] The study found the latter group were indifferent to the world and unresponsive to people who tried to hold and comfort them.

For those who lost their mothers during this critical period of development of social connection and emotion regulation, besides losing the mother, the surviving parent would be in grief for some time. If other family members or significant others cannot adopt the mother's role to help the child to regulate his or her

---

[42] Spitz, RA 1965, *The First Year of Life. A Psychoanalytic Study of Normal and Deviant Development of Object Relations*, International Universities Press, New York.

emotions, the child may learn to turn off his or her emotions and may have trouble maintaining attachments.

In reality, there are all kinds of mothers who cannot provide the conditions their children need for optimal development. There are mothers who encourage their children to be dependent, which can hinder development of abilities. There are mothers who do not provide care at all, and children are left on their own. There are immature mothers, who are self-centred and insensitive. There are unhappy and discontented mothers, who are plagued by self-doubt. There are insecure mothers who are moody. There are overburdened mothers, and even psychologically disturbed mothers.

The lack of a secure and supportive bond between the child and parent can have a serious negative long-term impact on the child. The early programming of nerve pathways that are formed under less than optimal conditions in the infantile brain will become established and consolidated in the child's developing brain. If a child only has one primary carer, which being an ill-equipped mother, there is a high risk that the child's feelings, thinking and behaviour will be influenced negatively.

If we are wired in our childhood to process stress ineffectively, then we tend to stay in the stress cycle. We were not aware of what was happening early in life because we are unaware of our unconscious emotions. Therefore, we are unable to understand them. That is the reason why we do not have any idea how to return to balance in an adaptive way. The longer we spent using the ineffective nerve pathways, the stronger it becomes.

However, we do not need perfect parents. One of the ways of avoiding the excessively one-sided developments that come from the ill-equipped mother is to have other carers in the family besides the mother. The child needs someone who can help him or her to achieve a sense of security, overcome fear and learn skills to process stress appropriately. This can serve as an alternative to the maladaptive nerve pathways learned from the ill-equipped mother.

As the child grows older, he or she may also have the opportunity to develop close emotional ties with people outside the immediate family, who may have a positive impact on their development.

The nerve pathways established through parent and child interactions, and the way to process stress can affect the child's self-regulation for a lifetime.

Our experience in the early years is the most important and has lasting influence on how we regulate stress.

Our stories

What guides us in all our decisions is not our consciousness. It is also not the knowledge that we have learned by rote; rather it is the experiences we have accumulated in our development. Historical events and past experience can influence our lives both positively and negatively.

The experiences we have had in the course of our lives become firmly anchored in our brains. Our experiences define our expectations and steer our attention in specific directions. Our experiences also determine the valuation we put on what we live through and how we react to the surroundings.

Experiences from our past have sneaked into our brains' unconscious regions and without us even realising, and they continue to influence our behaviour today. Even though the experiences might have been trivial, seemingly harmless, long forgotten and no longer valid today, they still continue to influence us.

The experiences we acquired are the most important and most valuable treasure a person possesses. We can use them not only for ourselves but also to pass them along to others. They do not become smaller, but get bigger and bigger as we use them and share them.

Events from our childhood and what we learn and do not learn can have a powerful effect on us. The reason is that at that time, we had not yet learned how to control our emotions or how to protect ourselves from events or interpret them correctly. A large part of our personality, our understanding of who we are and how we respond to others emotionally was still developing. Our unconscious emotional nerve pathways are aroused during different events and experiences, but our logical frontal lobes were still not fully developed.

As a result, these historical childhood events often continue to disproportionately influence our adult lives. For this reason, it is very important to understand our history and its effects on our responses and behaviours. For example, someone who grew up in a loving environment would interpret a hug differently from those who grew up in an abusive environment.

Some of our thinking and behavioural strategies have proved particularly successful or unsuccessful in our lives again and again. Consequently, we regard certain responses to be appropriate or inappropriate for solving future problems.

For those who had suffered from trauma, the events might disrupt how they view their own life stories. Instead of avoiding thinking about the trauma, one of the paths to healing is to seek assistance to reconstruct their stories coherently.

> What was the trauma?
>
> How did the trauma affect them?
>
> What were the consequences?

Alternatively, we can tell our stories to ourselves, even if nobody else is listening. When we can label our feelings, we have some mastery over them. This may help us to view the world as a more predictable and less frightening place.

Healing cannot occur until we are honest and have the courage to look below the surface of the stories that we told, identify the sadness and fear that underlie anger, the insecurity that expresses itself in arrogance, and the sense of meaninglessness behind most unhappiness.

Our stories and our experiences are important because it is our subjective evaluation of the knowledge built up in the memory about strategies for thinking and behaving.

The aim is to understand how the interpretation of our history affects us. Then, we can work out how to overcome the negative effects from our history, and make powerful changes in order to maximise our inner balance.

We are truly in control of our own stories. We can actively re-create our stories in ways that will help us function better.

Our aim is to understand our history, our interpretation and its effects on our responses and behaviours. Then, we can make changes and function better.

3 OUR MINDS AND HEARTS

Why some are unable to tell their stories?

Our childhoods leave us stories. Our experiences affect our feelings, our thoughts, our physiology and the interactive cycle goes on. We are often haunted by important relationships and experiences from the past that influence us unconsciously in the present. Our experiences affect our emotions and act as filters that influence how we interpret events that happened to us.

> *All sorrows can be borne if you put them into a story, or tell a story about them.*   Karen Blixen

What if a person can't tell a story about his or her sorrow?

There are stories that we never find a way to voice, because no one helps us find the words. When we cannot tell our story, our story tells us. We may then find ourselves acting in ways we don't understand. We may develop physical symptoms. Our mind does not generally send error messages in the form of conversation to us. Our mind recognises all the perceived threats from the outer and inner world and tries to compensate for them. However, when it is no longer capable of dealing with the fear and stress, it sends out alarms to us like the warning signals in a car dashboard. The message may be in a form that makes us physically or psychologically sick. It can just be a common symptom of insomnia, which frequently associates with underlying anxiety or depression or both. It is important to discover the cause because our body is trying to tell us something. However, many people just shrug their shoulders and try to continue with their usual activities. This can result in further damage.

Sometimes we may dream those stories. Sigmund Freud believed that recurring dreams, with a relatively unchanging structure, often contain memory fragments of early trauma. A person confided me with her troubles told me that she was often haunted by recurring dreams of her trauma. She was often awakened in terror. The basic structure of her dreams did not change. There were periods of time when she felt safe and the nightmares gradually occurred less frequently. At some point, her mental and emotional states improved, and she reported she did not have any nightmares for two years. However, when she was again confronted with a stressful situation, the same nightmares came back.

An infant's interaction with its mother is particularly important. In our early life (up to two years), mothers principally communicate nonverbally to reach the infant's right cerebral hemisphere, which generally processes nonverbal communication, visual-spatial and emotional perceptions and behaviour.

In young children (up to about two years old), the unconscious memories in which words are not required were developed before the conscious memories that are often language-based. Therefore, the interactions that take place from birth to the first two years of the infant's life mainly rely on the memory system that is generally unconscious.

A study that investigated the growth spurts, onset and rates of development of the human cerebral hemispheres, found that the left and right hemispheres developed at different onset times and rates, and that the timing of growth spurts overlapped the timing of major developmental stages described by Jean Piaget.[43]

Subsequent brain scan studies showed that the right brain hemisphere is dominant in human infants up to about two to three years of age.[44] Therefore, generally

---

[43] Thatcher, RW Walker, RA & Giudice, S 1987, 'Human cerebral hemispheres develop at different rates and ages'. *Science*, vol. 236(4805), pp.1110-3.

[44] Chiron, C Jambaque, I Nabbout, R Lounes, R Syrota, A & Dulac, O 1997, 'The right brain hemisphere is dominant in human infants', *Brain,* vol. 120, pp. 1057-65.

children up to about two years old are not proficient in verbalising about their experiences.

The critical growth spurt of the right cerebral hemisphere is from birth until the second year. However, the growth spurt of the left cerebral hemisphere starts around the second year of life. The left cerebral hemisphere generally processes the verbal-linguistic elements of speech, and tasks that require conscious processing. Hence, young children (up to about two years old) are generally right cerebral hemisphere dominant and rely heavily on the unconscious, emotional brain.

The memory we use when we describe what we did over the weekend and with whom is referred to as conscious memory. The conscious memory system begins to develop around two to three years of age. It recalls specific facts, events and episodes. It helps to organise our memories by time and place, which is also supported by language.

Our nonverbal interactions with others, and our emotional memories are part of our unconscious memory system, which also includes procedural memory such as learned motor skills, as in riding a bike.

Most memories are formed in the hippocampus (within the limbic system) and some are transferred to long-term storage in other brain areas. In young children, with the generation of new nerve cells, the transfer of the memory to the new nerve cells may weaken the connection with the memories stored. This may also explain why we rarely retain memories from when we were very young.

People who have been traumatised in their first two years of life would have few conscious memories of their traumas. This may serve to explain why it is difficult for them to tell their stories and sorrows.

Young children in the first two years of life can store traumatic events in the unconscious memory. However, they may not be able to recall. The unconscious memories are available to the conscious mind only as emotional reactions.

The unconscious memories of the trauma exist and are commonly evoked when people get into situations that are similar to the trauma. Such unconscious memories often seem to come 'out of the blue.' As the unconscious memories are not classified by time, place and context, the past seems to be real as in the present.

When we are dominated by thinking and actions that are unconscious, it is very difficult to interrupt and redirect them without special techniques. If we can transform unconscious memories that we are often not aware exist, into conscious memories that have a clear context, then it would be easier to recollect an experience as part of the past.

What we need is to be able to put events and our experiences into words. When we are able to put the events and experiences into words, we can re-transcribe the experiences from unconscious memory to conscious memory.

Some of us may be able to find the link through other senses rather than verbal or written communications. A person told me he was able to retrieve emotions through the sense of smell. It brought back the memory, which gave him a better understanding of why he disliked situations and places associated with certain smells.

The olfactory bulb (the structure that receives sensory input about odour) is considered part of the limbic system (emotional brain). The messages from the olfactory bulb are carried straight to the limbic system. It does not pass through other structures. This explains why the olfactory (smell) memories can take us further back than visual or auditory clues, which are believed to have to travel through the visual or auditory cortex.

The smell we encountered in the past was linked to the emotions we associated with the event at that time. Re-experiencing the smell may trigger the link and evoke the associated emotion of the event. This also applies to sight and sound. However, smell seems to be strongly associated with memory because the olfactory structures are closely linked to the amygdala and hippocampus, which are part of the limbic system. This is the reason why smell can create intense and instant emotional responses.

A person told me that he found some musical pieces to be helpful in terms of enabling him to access those emotions he would otherwise be unable to retrieve. I was very grateful and appreciative of his kindness, in which he gave me a CD of - *Adagio for Strings*,[45] the musical piece he valued very much.

I have received feedback from a few readers about my books published earlier. They told me they found the books helpful because they expressed feelings and experiences that they had held inside for a long time. They did not know how to communicate these feelings to themselves and others. They recognised them when they read certain passages in the books. The printed words talked to them. Reading the books helped them make a link to their past that they had never really examined before.

---

[45] *Adagio for Strings* was voted by the audiences of BBC Today in 2004 as the saddest music ever written. Composer - Samuel Barber (1910-81)

Once we gain a better understanding of the causes of our defensive habits or maladaptive reactive patterns, we can make use of our innately plastic brain to turn the ghosts into ancestors. As we work through them, they will not be haunting us anymore. Rather, they simply become part of our history.

When we are able to put our experiences into words, we can bring the unconscious into our conscious.

3 OUR MINDS AND HEARTS

Everything that happens leaves a trace

Not all of our mental processing happens at a conscious level. In fact, conscious thought makes up only a small proportion of our brain's functioning. Our mind and body are hearing the stories from the unconscious mind.

During the worst nightmare that we had, we found ourselves in great fear; our hearts were racing and we were sweating. We were so scared it could jolt us out of bed. It was just a dream, it was not real, nor conscious thought. Why did we react that way? We reacted that way because we were functioning from our emotional brain (the limbic system), which also stores our unconscious memories.

The conscious mind is what we know we know, and the memory is stored in the analytical brain (cerebral cortex).

The unconscious memory is primarily stored in the emotional brain (limbic system). It is what we don't know we know. The unconscious emotional memory from the past is actively infiltrating our current behaviour without us knowing. This explains why some of us suffer the consequences of a long-ago choice that we don't even realise was a choice. For example, we prefer people with certain attributes to be our partners.

Most of us would be adamant that we are not impacted by our childhood experiences. This is because unlike conscious thoughts, we can easily dig up the information. However, the link between the unconscious emotional memory and the conscious thought is not that obvious.

The unconscious mind is elusive as it is not easily under our control. It does not easily allow us to pick and choose the memories stored in the unconscious mind. All our experiences get stored in the unconscious mind. However, those experiences that are stored during times of distress may be misleading. The unconscious mind is the survival-vault of our mind; it can create sets of beliefs that can easily trick or dominate the conscious mind.

On the bottom layer, deeply anchored, we find attitudes and convictions adopted during our childhood, with not-so-clear traces regarding their associations with our conscious thoughts and feelings.

There is ample evidence to support that most of our behaviours or responses are products of underlying motivations, emotions, impulses and biases, which constitute Sigmund Freud's 'unconscious mind.' For example, if a child was treated like a princess, the childhood experience may be locked in her mind that the world owes her. This would have an influence on her behaviour in later life.

Conversely, if a person was abused or neglected in his or her childhood and was made to believe he or she was worthless; this unconscious emotional memory would shape his or her behaviour in later life. I once met a person with this unfortunate experience. When asked how she felt about herself, she believed she was worthless, though she was quite capable and intelligent. In fact, through observing the failures in her relationships and choice of partners, it is not difficult to see that she often chose men who were unkind because she thought she did not deserve the love of others.

Martin Seligman in the 1960s introduced the concept of learned helplessness. It is a state in which a person believes oneself to be stuck in a situation over which one has no control. The person also believes that any efforts to improve that situation are futile, so no such attempts are made. Actually, people with this perception tend to take little responsibility for their own decisions. They feel unworthy of respect, hence they do not demand that respect. Often, they find it difficult to express or communicate their inner experiences, and they feel isolated and lonely.

The examination of our unconscious will enable us to gain something we did not have before; we make ourselves a bit clearer to ourselves.

This is a true story [46] of a brave Melbourne woman, who has shared her story with the hope to help others find inner peace. The woman's parents were separated when she was a baby. Her mother worked two jobs from very early until very late to support themselves. Even at an early age, she had to walk to and from school on her own.

One summer afternoon in 1979, when the woman was five years old, she was late back home from school. Her older step-sister who attended secondary school at a different location was worried and asked why she was late, and where she had gone. The woman did not tell her step-sister what happened, nor was she able to talk to anyone because she lived in a dysfunctional family. The woman sometimes lived with her mother, other times with her father and step-mother or grandmother. What had happened to the woman stayed as a dark secret for a long period of time, and it seemed to have buried deeply inside, but it was not.

---

[46] Johnston, C 2013, 'Casting out the demons', *The Age Newspaper*, 22 Dec.

The woman married at 27 and had four children. When her daughter turned five in 2005, the same age as herself in 1979, the awful event just came back and hit her. All through the years, her unconscious mind had tried to bury the memories of the dark event. She was overprotective of her daughter and she ended up with severe depression.

In 2013, the woman picked up a newspaper and read about a notorious Melbourne paedophile, who had just killed himself in prison. The paedophile used to pick victims up in a car near their homes, and then he gave them money. The reference to being given money was the trigger. It set off alarm bells and repressed memories the woman had started to surface.

The woman told her husband that she might have been one of the paedophile's victims. She started to recall the event, in which a car pulled over and a man said to her, "Your mum said I had to take you home." She was only five years old, did not think it was unusual, and just hopped into the car. The woman could remember that the man abused her, and she remembered the car seat was cold. Then, the man gave her some money and drove her to a milk bar. She bought ice cream for herself and her sister, and had some change. But then she was lost; she could not find her way home. She ate the ice-cream and then sat on a bench and cried. Eventually, a passer-by drove her home.

The woman started seeing counsellors because what had happened affected her marriage. However, she felt that talking for an hour then to go back a week later and talk for another hour was like reopening the wounds.

She later found a counsellor, who was the one who started her healing. The counsellor had tears in his eyes and said to her, "I am really sorry about what happened to you." The incredible experience for the woman was finding another person who understood what she had been through and truly related to what she had experienced. That started her journey to control her disturbing memories and find her path to inner peace.

Being a mother has given her the strength of never wanting that to happen to her children and others' as well. The woman then reported the matter to the police. Investigations showed that the man who abused her was not the one reported in the newspaper who killed himself. The fact that the police took the matter seriously, showed compassion towards the woman in the process, and eventually came up with a suspect helped her find closure.

Anecdotal evidence shows that often it requires us to go through the very dark night of the soul in order to understand ourselves better.

The key is that the unconscious emotional memories are only accessible in times of stress. Also, rewiring of our nerve pathways has to be done through this portal during time of stress. This concept is supported by anecdotal findings that positive thoughts alone are not able to achieve this.

Freud conceived the existence of unconscious strategies or psychological defence mechanisms, which we use to cope with reality and to maintain our self-image.

We all have defence mechanisms (reactive patterns) that hide unbearable painful ideas, feelings and memories from conscious memories.

The 'immature' strategies serve to enable us to avoid the unbearable experiences. They also protect us from unacceptable or unbearable truths. Examples of maladaptive strategies include regression, repression, denial and the inferiority complex.

From the neuroscience perspective, if a nerve pathway is blocked, then the older pathway that is in place long before the established one would be used. This can be applied to explain Freud's concept of regression, in which one reverts to earlier stages of development rather than handling the stressful situation in a more mature manner.

What is not created through logic also cannot be removed by logic alone. Healing is about using our emotions to track down those memories, and bring them to the attention of the analytical mind.

The aim is to develop new nerve pathways that facilitate mature responses such as patience, humility, acceptance, gratitude, altruism and forgiveness.

These mature defence mechanisms will keep us from over-reacting and will enable us to optimise our relationships and functioning in the community.

Everything that happens to us leaves a trace in our unconscious mind, which can affect our behaviour.

3 OUR MINDS AND HEARTS

Healing through rewiring our brain pathways

Many past stresses are encoded in the nerve pathways that lie behind the conscious. The brain remembers better in highly stressed states; whatever happens in this state can quickly become a nerve pathway. If these emotions and responses are repeated and reinforced, the nerve pathway will be established, just like a thick rope that grows from thin filaments one by one.

In order to heal the emotional stress we need to access and rewire these emotional nerve pathways. The nerve pathways that are most problematic are usually encoded early in life, of which we are not aware of the past link. We are not consciously aware of the nerve pathways that encoded the unconscious emotional memories, which were wired together a long time ago.

The unconscious memory does not have time. This explains why we feel the emotions as if they are happening now. Past and present are being mixed up together. When we are dominated by thinking and actions that are unconscious, it is very difficult to interrupt and redirect without special techniques.

According to Freud, sometimes we unconsciously transfer ways of perceiving from the past onto the present. This way, we are reliving the past, rather than remembering the past. This occurs when we view others in a distorted way without being aware of doing so; that can often lead us into difficulty. For example, we might regard our supervisor at work as the person who had treated us badly in our critical psychological period. Freud referred this as 'transference.' It is through understanding the transference that allows us to improve our relationships.

How to find the emotional nerve pathway that we want to change?

The emotional nerve pathways are often laid down in the unconscious; therefore it requires creative ways to access them.

One way is to begin to tell the story of our lives. When telling one incident that triggers a stress response, we are closer to finding the emotional nerve pathway that we want to change. We need to be able to put events and our experiences into words.

Alternatively, we can arouse an emotional pathway by drawing upon an emotional experience similar to the one that encoded it.

What is the window for change?

Usually it is when we are stressed (but not extremely), which is the time we can find the nerve pathway; it is also the window for us to rewire the nerve pathway. This is because the synapses (the links between nerve cells) will become fluid and open to change when we are stressed.

It will not work when we are extremely stressed; that is when all the symptoms of stress are fully blown. This is because we are focussed on survival in extremely stressful situations; we are overwhelmed and confused. At that stage, we are functioning at a lower level. We cannot even connect with ourselves.

When we are just a little stressed, it will not work either. This is because the brain can restore the balance and settle back to our usual state.

What do we do when the nerve pathway is open?

Once it is open, the key is to respond to it differently, and feed it a different experience.

The emotional nerve pathways that are most problematic are usually encoded early in life. When they are activated repeatedly, they are like thick ropes and it may take a lot of effort to weaken them.

The same principle in forming the emotional nerve pathway applies to unlearning. Rewiring our nerve pathway requires repeated exercises. This is crucial, as it serves to 'delink' the pathways. Based on the principle that nerve cells that fire apart wire apart, when we respond differently, a new nerve pathway is formed. When we do it often enough, this new nerve pathway is gradually reinforced.

An example of a technique to achieve emotional transformation is to find a substitute situation or imagery that vividly captures the essence and the emotions of the event causing the unwanted emotional responses.

Then, we need to evoke the exact opposite of our current maladaptive response that will work to our advantage. The purpose is to disconnect from the negative emotions we invoke when we enter the substitute situation or imagery. Step by step, we aim to weaken, whittle away or break the current emotional nerve pathway and replace it with one that is more constructive.

Healing

The objective is to rewire our emotional connections so we have different emotional reactions. New ways of relating have to be learned through wiring new nerve cells together, and old ways of responding have to be unlearned to weaken the nerve linkage.

If we can achieve the breakthrough through repetition, it will lead to long-term changes in our behaviours and responses. Then, healing will occur.

In the end, we can find forgiveness for neglect or abuse. We can find peace about a divorce or job loss. We can identify the roots of an addiction. We can acquire a clear picture of our lives so far.

Psychotherapy is a biological treatment

Eric Kandel (Year 2000 Nobel laureate in Physiology or Medicine) argues that psychotherapy can result in detectable changes in the brain. The change is made through altering the synaptic connections, resulting in structural changes that also alter the anatomical pattern of interconnections between nerve cells of the brain. Therefore, psychotherapy is considered a biological treatment that produces lasting and detectable physical changes in our brain, as learning does.

Review of studies since the 1990s regarding changes in the brain after different psychotherapies for common psychological disorders suggest that they change brain function.[47] Brain scan studies showed that structures in the limbic system (emotional brain) are abnormally activated in anxious patients with panic disorder. However, there are reduced activities following psychotherapy. The finding suggests psychotherapy might have some influence on the prefrontal lobe, which plays a role in moderating emotions.[48]

Plastic changes in the human brain are difficult to study. Most studies investigated the changes on the whole brain systems level (through measuring changes in brain blood flow or metabolisms). There are relatively fewer studies that measured molecular level changes after psychotherapy.[49,50] Studies that investigated the use of

---

[47] Karlsson, H 2011, 'How psychotherapy changes the brain', Aug, *Psychiatric Times*, Special reports.

[48] Beutel, ME Stark, R Pan, H Sibersweig, D & Dietrich, S 2010, 'Changes of brain activation pre post short-term psychodynamic inpatient psychotherapy: an fMRI study of panic disorder patients', *Psychiatry Research*, vol. 184, pp.96-104.

[49] Lehto, SM Tolmunen, T Joensuu, M Saarinen, PI Valkonen-Korhonen, M Vanninen, R Ahola, P Tiihonen, Kuikka, J & Lehtonen J 2008, 'Changes in midbrain serotonin transporter availability in atypically depressed subjects after one year of psychotherapy', *Progress in Neuropsychopharmacological Biological Psychiatry*, vol. 32, pp.229-37.

psychotherapy and anti-depressants showed that in depression, psychotherapy and pharmacotherapy seem to operate through different nerve pathways. It appears that psychotherapy takes the more 'top-down' approach through regulation of hyperactivity of structures in the limbic system by the prefrontal control system. Whereas, medications take the more 'bottom-up' approach that operate more directly on the structures of the limbic system.[51] The focus of recent developments is on using scans of brain activity to predict which group of patients respond better with either approach or a combination.[52]

---

[50] Hirvonen, J Hietala, J Kajander, J Markkula, J Rasi-Hakala, H Salminen, JK Nagren, K Aalto, S & Karlsson, H 2011, 'Effects of antidepressant drug treatment and psychotherapy on striatal and thalamic dopamine D2/3 receptors in major depressive disorder studied with [11C]raclopride PET', *Journal of Psychopharmacology*, vol. 25(10), pp.1329-36

[51] Linden, DEJ 2006, 'How psychotherapy changes the brain – the contribution of functional neuroimaging', *Molecular Psychiatry*, vol. 11, pp.528-38.

[52] McGrath, CL Kelley, ME Holtzheimer, PE Dunlop, BW, Craighead, WE Franco, AR Craddock, RC, & Mayberg, H 2013, 'Toward a neuroimaging treatment selection biomarker for major depressive disorder', *JAMA Psychiatry*, vol. 70(8), pp.821-9.

As more evidence from functional and metabolic imaging becomes available, more detailed models of the nerve pathways of psychotherapy effects can be constructed. It seems Sigmund Freud had the vision a century ago that the effects of psychotherapy could be established through neurobiological science eventually.

We must recollect that all our provisional ideas in psychology will presumably one day be based on an organic substructure. - Sigmund Freud -On Narcissism 1914

The deficiencies in our description would probably vanish if we were already in a position to replace the psychological terms with physiological or chemical ones....We may expect physiology and chemistry to give the most surprising information and we cannot guess what answers it will return in a few dozen years of questions we have put to it. They may be of a kind that will blow away the whole of our artificial structure of hypothesis. - Sigmund Freud - Beyond the Pleasure Principle 1922

The purpose of psychoanalysis is to help the patient to bring unconscious thoughts and feelings that had been repressed to the surface of the conscious mind. This way the patient can learn more about their genuine thoughts and feelings about life situations that cause conflicts and tension, but are not self-evident.

When the experiences and events are in the conscious memory, the psychoanalyst will try to help the patients point out the emotions, triggers and how they influenced their mental and bodily states. The aim is to enable the patients to spot the triggers and emotions themselves eventually. Then, the patients can try to re-write or re-interpret their stories and memories rather than relive the traumatic memories or experiences.

According to Sigmund Freud, the positive sense of closeness developed between the patient and the analyst is the element that promotes the cure. From the neuroscience perspective, the positive bond appears to facilitate the change by triggering unlearning of the existing nerve pathways. This gives the patients the opportunity to look at the unconscious emotions and their current relationships, so as to develop insight and make positive changes.

One time, a woman confided and shared her unfortunate experiences with me, in which she talked about her repeated relationship failures over decades. The woman had not worked through her feelings of low self-worth, and there was an emptiness inside her which made her feel needy for love. She needed others to create her self-worth and fill her with love. Emotionally she was still at an earlier stage of development. As her dependency needs were not fulfilled, her motive was to expect others to give her love. Her hunger for love was controlled by her unconscious desire for love that she lacked in her childhood. The pronounced character trait, fear of loneliness and her compulsion to fill her emptiness was not genetically predetermined but learned from her childhood experiences. She was not making decisions based on sound rational judgment. She would accept partners who were not suitable. The outcomes meant that the relationships ended after a short time period and she then started the cycle again. That pattern led to ever-deepening frustration as her relationships never worked. It was just going from one disappointment to another.

The type of relationship she searched for was not mature. Unconsciously, she was searching for people whom she would like to attach to, someone who she thought would like to fill her with love, look after her and give her everything she wanted. Often, the persons she chose would have the same expectations. In a relationship where both parties just want to receive love rather than sharing and giving love, sooner or later the relationship will end in disappointment. This is because each party will blame the other for not giving each other enough. In some ways, the extent to which we need to receive love is a test of the level of our emotional development. The reactive pattern that the woman used had been reinforced through repeated applications for many years. At one point, I thought it was the right time to put forward to her that the main reason her relationships were not successful was because she had not learned how to love. I explained to the woman that it was because she had never been loved truly. Working through the reactive pattern, this had helped the woman to realise her mind was fused with the idea that she was worthless. In the process, the woman was able to connect her desperate searching for love with its true trigger, the absence of loving parents in her early years. Making this connection, and also realising that she was no longer a helpless child, she felt less overwhelmed. This enabled her to unwire the connection and the maladaptive patterns of behaviour. Subsequently, the woman understood her behavioural pattern and what was influencing her. She no longer blamed her partners and others for her misfortunes. The woman understood she did not choose her partners wisely and she was partly responsible for all the failed relationships. She was able to face her pain and the reality.

When we understand why we react to some events as though they are major losses, such as over-reaction to a breakup, that is the insight that we need to develop; the epiphany of our lives. Once we understand the causes of our defensive habits, and identify the pattern of our maladaptive behaviour, then we can change and live a better life.

We can change our lives by understanding ourselves and having control over it.

3 OUR MINDS AND HEARTS

How do we see the world?

On a heath, King Lear asked Gloucester, "How do you see the world?" Gloucester, who was blind, answered, "I see it feelingly." (*Shakespeare's King Lear*)

The human brain acts as a filter, so we often learn only those things that reinforce our existing views. Our brain analyses the sensory information, filters it, sorts it, assesses it, deletes some of it, fills in the gap, categorises the information, interprets and ascribes meaning to it and then forms a plan of action.

When new information conflicts with our well-established preconceptions, we put on our emotional blinkers and mental tinted glasses. The information that conflicts with our view is ignored, denied or interpreted in such a way that our beliefs and worldviews remain intact. We are more or less sensitised to certain perceptions. We are also able to dumb down our senses consciously at first, and later unconsciously. This way, we can perceive and relate to certain specific phenomena better and faster than others. This is an efficient way to process information. Sometimes, it is necessary. However, if we do not keep an open mind, we shall be closed to learning new things. Eventually, it can lead to rigid thinking and ignorance. For example, homosexuality is an inborn trait, but there are groups that consider it a sin or lifestyle choice.

What we see might not be a true picture of the actual nature of the external world; it is the picture we are capable of making about the world with our limitations. For instance, we can only see light of certain wavelengths, but we can learn enough about the world to survive. We are seldom conscious of the fact that we do not perceive the whole range of signals from our inner world, for example changes in blood sugar and concentration of oxygen level. They do not usually get noticed unless they seriously affect our wellbeing.

Our perceptions often slant toward a distorted reality. We unconsciously link a number of things or characteristics together based on what we think we know to be true. We often exaggerate differences and minimise others. In the end, not everyone sees the world with an open mind.

Sound comes to us; noise we come upon.
Hillel Schwartz

We tend to go to great lengths to continue to see the world through the perspectives we are familiar with and will do whatever we deem necessary to protect what we believe to be real.

We see the world not as it is, but as we are.
Anaïs Nin

To see the world with an open mind and open heart actually goes against our habitual pattern. However, we need to do this in order to get a clear and impartial way of seeing.

We can see more with open hearts and open minds.

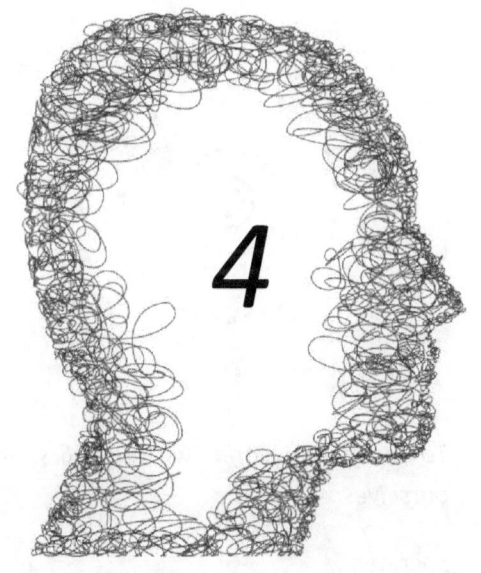

Learn more and do more

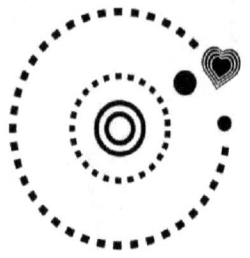

To move the world, we must first move ourselves.

Socrates

4 LEARN MORE AND DO MORE

We do not know anything

Why did we come to exist 13.7 billion years ago in the Big Bang? Unknowing remains an essential part of the human condition. We are all a 'work in progress'; our understanding of ourselves, the universe and the divine is also a 'work in progress'.

Most of our behaviours are driven by experiences, history, culture, unconscious desires and motives, of which we are only vaguely aware. It is possible that much of what we know is untrue.

Advances in modern science are pushing back the frontiers of ignorance. It will soon lay bare the last secrets of the universe. Science is by its nature progressive; it continually breaks new ground. Existing theory is disproved and continuously surpassed with new findings.

However, knowledge of arts and humanities does not advance in this way. Therefore, after hundreds of years we keep asking the same question – what is the purpose of life, what is truth, and what is happiness? It rarely arrives at a definitive answer, because there are no definitive answers to these questions. Each generation has to start over and find solutions that speak directly to its unique circumstances.

Philosophers today still discuss the same issues that preoccupied Plato (429–347 BCE) and others a long time ago.

When we are pushed to the very limit of what is knowable, we step outside our selves and tip over into transcendence. However, we can never understand transcendence because it is transcendent.

Similarly, the supernatural power or divine lies beyond the reach of our five senses, and is therefore incapable of providing definitive proof.

Religion is at its best when it helps us ask questions and holds us in a state of wonder. At its worst, it tries to answer them authoritatively and dogmatically. Omniscience, the capacity to know everything that there is to know, which forms the foundation of the common religions, can lead us to form views of others coloured by our own fears. The omniscient point of view generally is coloured by prejudice, which attempts to serve one's agenda.

Socrates (469-399 BCE) believed that wisdom was not about accumulating information and reaching hard and fast conclusions. He insisted the only reason he could be considered wise was because he knew that he knew nothing at all.

> *I am wiser than this man; it is likely that neither of us knows anything worthwhile, but he thinks he knows something when he does not, whereas when I do not know, neither do I think that I know; so I am likely to be wiser than he to this small extent, that I do not think about what I do not know.*

When the intellectual foundation of our lives has been radically undermined, that is the moment we realise the profundity of our ignorance. We shall realise that our views of others are perpetually clouded by our prejudices, opinions, needs and desires.

It is egotism that makes us identify with one opinion rather than another. We become quarrelsome and unkind. We think we have a duty to change others to suit ourselves. Actually, we do not understand nor truly respect others.

> *An unenlightened person is like a frog in a well who mistakes the tiny patch of sky he can see for the whole; but once he has seen the sky's immensity, his perspective is changed forever.*
>
> Zhuangzi – Chinese philosopher (369-286 BCE)

The capability of the human brain enables us to thrive on solving mysteries as a way of removing doubts and pushing our society forward from the darkness of ignorance into the light of knowledge.

Our aim is to recognise and appreciate the unknown and unknowable. Our know-it-all world doesn't know it all.

Humility in the face of mystery is a good thing. Humility helps us to become more sensitive to the over-confident assertions of certainty in ourselves.

> *Knowledge is what we learn from others.*
>
> *Wisdom is discovered within ourselves.*
>
> *It comes from our hearts and minds.*

The most important philosophical truth is that we do not know anything.

We need teachers

Most of us know what to do, we just need models of how those who have gone before us have reified their beliefs.

We need other people to teach us to speak a new language. It is difficult to learn to swim through reading a book, or fly an airplane by reading the manual. Therefore, we need to seek guidance from competent practitioners. We need someone to teach us.

For our personal and or spiritual development, we need someone important to us, with whom we form a close emotional relationship. None of us will take advice from strangers, and often not even from our parents or spouses.

We do not like our teachers because they are good-looking, but because they are as they are. There are things our teachers can do and know, and also things they cannot do and do not know. We need to find someone whom we respect; someone who is able to understand us better, someone who asks uncomfortable questions, and encourages us to look below the surface of our lives.

We need to be able to identify those among us who are qualified to teach us. They must be wise and devoted to the principles of kindness. We need people who think, feel and behave differently than we do. If we follow the wrong people, we can only blame ourselves.

However, there are things our friends cannot teach us, such as patience, tolerance and forbearance. Only our enemies can teach us to be patient, practise tolerance and forbearance. There are a lot of situations in which our enemies had at some time been our teachers. Our enemies are the best teachers of altruism. They provide us with opportunities to develop compassion. We develop love for others through extending it to those whose motivation is to harm us.

> *I have learned silence from the talkative, toleration from the intolerant, and kindness from the unkind; yet strange, I am ungrateful to these teachers.*   Kahlil Gibran

Spiritual advancements require enemies as instigators, and learning from tragedy. We thank our enemies for blocking our wishes, and we shall respond with gratitude. The experience of suffering will enable us to empathise with the pain of others and the urge to do something for them.

Difficult periods offer us a chance to develop inner strength, determination and courage to face the problem. It also allows us to get closer to reality, peeling off all pretensions. We try to avoid situations that are painful. However, sometimes it is unavoidable. When it comes, we shall receive it not as a burden but as something that can assist us.

Adversity is a bittersweet fruit. We cannot be a master sailor on calm seas. Often, adversity results from being in the wrong place at the wrong time, or being born into a less than perfect family.

There are also adversities that we bring upon ourselves by making mistakes through error of judgment or risk taking. Anyway, it is all about what lesson we can take from it, and how we can do things differently to prevent it from happening in the future.

> *Learn from the mistakes of others. You can't live long enough to make them all yourself.*
> Eleanor Roosevelt

Even when life is at its most painful, it has meaning and value. It can be painful, but if we can move beyond the initial feeling there is some value in the experience. Some good can come of it.

Hopefully, as we age, we become wiser through avoiding trouble and dealing better with it when it arises. If we can draw connections between what we have learned from the crises, we can work out a systematic way of dealing with them.

We can be teachers for others as well. There are people who are walking along the edge of a cliff, and each step brings them closer to misery. The greatest help we can offer is to lead them to true happiness, the ultimate aim of our lives. Teaching others to understand themselves and tame their minds can liberate them from suffering.

How do we find our teachers?

> *When the student is ready, the teacher will appear.*   Zen saying

In order to find our teachers, we may have to make ourselves ready to be an authentic student. Sometimes, we may need to be away from our usual lives in order to meet our teachers.

> *Sometimes when we travel to faraway places, we may meet a person whom we can connect profoundly.*[53]   Yu Qiuyu

---

[53] Yu Qiuyu (余秋雨), contemporary Chinese cultural and literary scholar.「走得遠了，也許會遇到一個人，出現在你與高山流水之間，短短幾句話，使你大惊失色，引為终生莫逆。」

I once met a traveller from overseas for the first time and we had a very enjoyable discussion about the philosophy of life. A few days later, I received a note from her. I was humbled when I read the message, "A single conversation with a wise man is worth ten year's study."[54]

The best teachers teach from the heart, not from the books.

---

[54] Chinese saying 「聽君一席話，勝讀十年書。」

很久以前，有個窮秀才進京趕考。他只顧趕路，錯過了宿頭。眼看天色已晚，他心裡非常著急。正在這時，一個屠夫走過來，邀他到自己家裡去。屠夫與秀才談得很投機。於是屠夫隨口問秀才說：「先生，萬物都有雌雄，那麼，大海裡的水哪是雌，哪是雄？高山上的樹木哪是公，哪是母？」秀才一下被問呆了，只好向屠夫請教。屠夫說：「海水有波有浪，波為雌，浪為雄，因為雄的總是強些。」秀才聽了連連點頭，又問：「那公樹母樹呢？」屠夫說：「公樹就是松樹，『松』字不是有個公字嗎？梅花樹是母樹，因為『梅』字裡有個『母』字。」秀才聞言，恍然大悟。秀才到了京城後，進了考場，把卷子打開一看，巧極了，皇上出的題，正是屠夫說給他的雌水雄水、公樹母樹之說；很多秀才看著題目，兩眼發呆，只有這個秀才不假思索，一揮而就。不久，秀才被點為狀元。他特地回到屠夫家，奉上厚禮，還親筆寫了地塊匾送給屠夫，上面題的是「聽君一席話，勝讀十年書。」

Mindfulness

We often miss the subtleties in our lives. We drive through the same neighbourhood for years. Then one day we notice the most beautiful magnolia tree; it has always been there, but we have never really seen it. We have been too busy hurrying through our lives.

A successful business-woman has accumulated enough wealth after years of hard work. One day, she stopped to reflect and decided she would like to participate in charity work. The woman admired Mother Teresa's work with the poor in Calcutta. Therefore, she wrote to Mother Teresa and offered funds and herself to be involved.

However, Mother Teresa replied, "Stay where you are. Find your own Calcutta. Find the sick, the suffering and the lonely right there where you are — in your own homes and in your own families, in your workplaces and in your schools. You can find Calcutta all over the world, if you have the eyes to see. Everywhere, wherever you go, you find people who are unwanted, unloved, uncared for, just rejected by society — completely forgotten, completely left alone."

The following is a Zen story about focusing on the present.

A tiger sprang from the bush to chase a man walking in the wild. The man knew he had limited choice, so he jumped off the nearby cliff. Fortunately, he was able to hold onto a vine, dangling a few hundred metres above the ground. When he looked down, he saw another tiger waiting below. The man looked up again and saw the tiger that chased him was prowling the edge of the cliff. The man then held on as tight as he could, but he realised the vine could break any time. Instead of worrying about the tigers above and below, he stayed calm and looked around. He noticed a wild berry growing from a crevice in the side of the cliff. He reached and picked the wild berry and ate it. That was the sweetest wild berry he had ever tasted.

The sweetness of life symbolised by the wild berry is always there. However, we might not notice, enjoy and value it. Regarding the tigers, they are just part of life.

With mindfulness, we use our new analytical brain to step back and become aware of the instinctive, selfish and automatic mental processes generated from our emotional brain, which are the cause of so much pain. We tend to assume other people are the cause of our pain. However, with mindfulness we learn how often the real cause of our suffering is the primitive negative emotions that reside within us. We need to detach ourselves from our egos by observing the way our minds work.

Using functional magnetic resonance imaging and electroencephalographic technology, neuroscientist Richard Davidson found that people who tend to have negative emotional style have heightened activity in the right prefrontal cortex. People who tend to have positive emotional style and are enthusiastic about life have heightened activity in the left prefrontal cortex. Most people's mood states fluctuate depending on what is going on in their lives. However, when we are in resting state, we tend to revert to our baseline emotional style; some tilt toward the left, and some tilt toward the right. Davidson's findings provide empirical support to the concept that suggests each of us has a biologically determined baseline emotional style.

Davidson further demonstrated that we can change our baseline emotional activity style through continuous practice of mindfulness meditation that focuses on here-and-now.[55] The mindfulness meditation program serves

---

[55] Davidson, RJ Kabat-Zinn, J Schumacher, J Rosenkranz, M Muller, D Santorelli, SF Urbanowski, F Harrington, A Bonus K & Sheridan JF 2003, 'Alterations in brain and immune function produced by mindfulness meditation', *Psychosomatic Medicine*, vol. 65(4), pp. 564-70.

to enable us to detach from our ruminations and negative moods. The change in the emotional state can be attributed to the shift in attention and focus, which contributes to the activation of the left prefrontal cortex. From the neuroscience perspective, the more often we activate the nerve pathways that induce a positive emotional state, the easier it will be to induce that again and the more likely it will become a stable emotional style.

Richard Davidson conducted a research study that investigated brain activities in advanced practitioners of meditation.[56] The study showed that the nerve connections between regions of the prefrontal cortex and the amygdala provide the link to moderate negative emotions. Specifically, the left prefrontal cortex has the capacity to inhibit the overactivity of the amygdala, which is most active when people are distressed, anxious or angry. As part of the study, a Tibetan monk (Matthieu Ricard) was recorded to have excessive activity in his left prefrontal cortex.[57] The finding

---

[56] Davidson, RJ & Lutz, A 2008, 'Buddha's brain: neuroplasticity and meditation,' *IEEE Signal Processing Magazine*, vol. 25(1), pp.174-6.

[57] Bates, C 2012, 'Is this the world's happiest man? Brain scan reveals French Monk has abnormally large capacity for joy – thanks to meditation.' *Daily Mail Australia*, 31 Oct.

suggested that the monk had an abnormally large capacity for happiness and a reduced tendency towards negativity.

Different types of meditation can produce different changes in the brain. Mindfulness-based meditation can also be applied to assist people to manage their chronic illnesses such as chronic low back pain, which can get worse with stress.[58] The aim of future studies is to acquire better understanding of how different nerve circuits are used during meditation to produce mental and behavioural changes that promote wellbeing.

If we attend to all the information we receive, we shall be distracted by the unimportant information. Even if we are able to attend to all the information, we shall become exhausted and stressed. Being mindful can help us distinguish what is really important and what is background noise. It allows us to see the larger picture of life, and to direct our attention to what is important in our lives.

The cause of our pain and suffering is actually within us. Our horizon shrinks when we are engrossed in thoughts

---

[58] Morone, NE Greco, CM Weiner, DK 2008, 'Mindfulness meditation for the treatment of chronic low back pain in older adults: A randomized controlled pilot study', *Pain*, vol. 134(3), pp. 310-9.

of anger, hatred, envy, resentment and disgust. These primitive negative emotions make us unhappy. When the negative emotions are dominant, our logical thinking diminishes and we cannot respond and behave appropriately and kindly. When we are angry, we exaggerate others' shortcomings.

Through mindfulness, we become aware of how suddenly these impulses arise in response to stimuli that make us irrationally angry, hostile, greedy, frightened and possessive. We do not need to be overly distressed by our discoveries, but rather understand that this is what we acquire from our nature. These strong instinctual emotions have their function in primitive lives. However, in civilised societies, our goal is to control them in order to function well. Through a calm and dispassionate appraisal of our behaviour, we can become fully aware that our judgments are often biased and dependent on a passing emotion. Often, it is our endless self-preoccupation that brings us into conflict with people who seem to get in our way.

The past is unchangeable and the future is unpredictable; we can only have some control of our lives right now. To enter a different path, we need to pay more attention to the present rather than the past.

The aim of mindfulness is to use our mental energies in a more productive and positive way.

4 LEARN MORE AND DO MORE

Say no to past conditioning

Most of us lead an existence where we are more or less conditioned beings, a bit like machines. Our genes and our thoughts determine us. The experiences of the past also pulse into the present moment and affect our responses.

Some of us might live in conditions that did not allow our genetic potential to develop. Some of us might not have the opportunities to receive training to fully utilise ourselves.

However, there are those who are lucky enough to grow up in a world where they have the opportunity to maximise their genetic potential. They are also able to develop throughout their whole lives.

The majority of us would be less than fortunate in our childhood to actually find ourselves in such conditions. Therefore, we shall have to deal with the traces left in our brains by the less than ideal or outright inadequate developmental conditions.

Significant experiences acquired in early childhood and youth often led to stabilisation of certain nerve pathways. The established nerve pathways serve as templates that facilitate our perceptions and experiences later in life. This accounts for how we feel, think and behave in certain situations.

The nerve pathways and defective response patterns that were put into place during the early phase of our development are to a certain extent correctable during the adult phase.

In order to break down this kind of programming later in life, we first have to acknowledge the existing established nerve pathways. This is what Sigmund Freud refers to as making the unconscious conscious.

> *Until you make the unconscious conscious, it will direct your life and you will call it fate.*
>
> Carl G. Jung

Second, we have to address the imbalance that restricts the effective use of our brain's potential. We have to use all our senses and learn to grasp and evaluate reality. We need to be detached from our perceptions and beliefs, and find a balance between our emotions and rationality.

Third, we need to take in new concepts and perceptions from the inner and outer worlds, and integrate them to form a comprehensive picture of outer and inner reality.

The goal is not to continue to have our perceptual capacity determined by those circumstances that always compel us to see, feel and understand in a particular way. We need to allow the new external environment to enter us, and actively connect the new information with all the other information and beliefs that are in us from the past.

Our brain has the capacity to break down already established nerve pathways, rewire and restructure them - this includes beliefs and emotional patterns.

We can become more by starting to say no to past conditioning.

We are free only when choices are made without past conditioning.

4 LEARN MORE, DO MORE

Whom to marry?

Most people want to be in a committed relationship. For young adults, the most important choice of their lives is the choice of life partner. This is indeed the most important life choice. It is even more important than our careers.

Statistics show that a lot of people recognise their errors. Besides the heartbreak and divorce, they are paying for their mistakes for a large part of their lives. There are those who are still married, but deep down they know they have made a mistake. Children and finances are the common reasons that impel them to remain together. For those who failed the first time, it is important to learn from the first painful process before prematurely jumping into the next relationship.

Why it is so hard to get it right?

Love can make us blind. We see in our new love all the things we have been dreaming of, and we believe it is true. We believe the person we just met is the one who will love us as we are, grow together with us and share our dreams. It is the classic example of the emotional brain paying little heed to logic. For instance, there are people who fall in love with someone who did not want to or could not deliver the bare minimum of loyalty, trustworthy and kindness towards a mature relationship.

It requires wisdom to make good decisions about other people. We do not receive adequate training by the time we need to make a choice. We are poorly prepared to make the choice of selecting the person with whom we expect to spend the rest of our lives. Sometimes, we are not able to tell the good guys from the bad guys. Parents do not train their children to develop the ability to discern those traits of character that contribute to a satisfying and enduring relationship. Perhaps parents do not have the knowledge themselves. Anyway, young adults would probably not accept advice from their parents. They prefer to take advice from their peers who are equally not equipped.

Basically, we do not know what we base our decisions about whom to marry. We do not understand ourselves well. We do not know who we are looking for. Some just follow the simplistic way to select a partner. For example, men tend to draw to women who are physically attractive. Often, this may not work out well.

We are not able to predict what the other person will be like in 10 or 20 years. The person we married may change over time and become someone we no longer love. Perhaps, we may change into someone who is unkind. We have to understand we are all imperfect. We have to ask ourselves the question, how can two imperfect people form a perfect relationship?

We are not knowledgeable about what we can reasonably expect from others. For many of us who had good parents, the memory of unconditional acceptance lives on in our unconscious as our deepest longing. But it is difficult to find another person who will love us just as we are, so we learn to settle for something less. How much less is at the heart of most marital difficulties. It might be the fault of our mothers, who loved us so much that no other adult can match us.

On the contrary, those who did not have enough parental love will grow up as children looking for that unconditional love that few spouses are willing or able to provide. This unrealistic expectation can be the cause of failure or frustration.

We have to ask ourselves, is our decision based on whether someone is kind, loving, responsible, trustworthy, and having shared values?

Marriage as an institution is not failing us; it is just that we are not prepared, nor do we have realistic expectations.

Giving our heart to another is the most important decision in our lives.

4 LEARN MORE AND DO MORE

A relationship is a spiritual path

To have loved makes life much richer. Being married is a great opportunity to get to know ourselves through knowing another person.

A relationship is a spiritual path. Nothing in the world will push us to grow more. There is nothing more revealing, humiliating, uplifting and humbling than being in an intimate relationship with another human being. What better place to do that than in open exchange with our partner?

Loving deeply is an uplifting experience, yet it can be equally treacherous. The person we are with will serve as the mirror of who we are inside. All our walls of protection are going to be challenged to come down, and who we truly are is going to be called out one way or another. Our partner is also striving to become aware of his or her shadow side as well.

In order to open ourselves up in this way and to love, we have to take risks. It takes courage to allow ourselves to be truly known by another human being. This is essential for us to merge with another person and blend our lives with theirs.

Intimacy is built on what is authentic and real, being as we truly are. To be in a true relationship with another human being, we must construct our relationship on what is real.

However, some people choose not to take the risk. Perhaps life has been too painful because those who were supposed to love us betrayed our trust. This is illustrated vividly in a television documentary in the *Australian Stories*.[59] Richard Farleigh is an Australia-born multi-millionaire entrepreneur, who by his mid-thirties made his fortune from the financial market. He then settled in London. In his fifties, Richard was back in Australia to retrace his past, which he said continued to haunt him in ways he could not control. Richard has an unusual background, being one of 11 children from a violent, alcoholic and itinerant sheep shearer. His mother was also an alcoholic and the children were abused and neglected. Subsequently, Richard and several of his older siblings were placed in orphanages and fostered. Richard lost contact with his siblings because most of them adopted their foster families'

---

[59] "Australian Stories - There but for fortune", 2013, television program, Australian Broadcasting Corporation, 22 July.

names. A sister-in-law of Richard's, who worked in the Birth Registry, noted an application for a change of name that might be one of her husband's brothers. Further investigation subsequently led to the reunion of Richard with his siblings. Richard and his siblings then started to trace their parents, and subsequently learned their parents had died some years earlier. They visited their parents' graveyards to lay them to rest. When asked how they felt, they said they did not miss their parents at all.

Richard said the impact of his childhood had affected his marriage. He was divorced from his wife. To Richard, as much as he loved his wife, it was still difficult to put away all the things that had happened in the past. Richard was like most of those who did not have a secure and close emotional relationship with their parents. Richard had not learned how to express his emotions. He built a hard shell around himself and hid his feelings. That made relating to people at an intimate level particularly difficult. It is certainly not the way that enables an intimate relationship to grow.

In order to break through the barrier, one must be given the opportunity to enter into close emotional relationships with other people. A close emotional relationship is the only way one could gain experience that close emotional relationships could provide security. Under the realm of security, it is possible to let one's own world of feeling merge with another person's. This will enable one to learn it is not dangerous, but rather enriching to put oneself in another person's place and be able to feel what is going on.

What Richard needs is to learn to find out why his partner is infuriated by his aloofness. Probably, he needs to discover the part of himself that had been linked to his parents. Finding that part of himself that had once loved another person deeply may allow him to again truly love another person.

We need to transcend the fears we have. Our fears are there for our survival. When threats to our survival are no longer present, it is beneficial to remove the crusty protective shell over our hearts gradually.

Those who are afraid of closeness equate it with vulnerability and loss of control. When fear and mistrust keep growing, eventually interactions become unkind, critical, controlling and alienating. Eventually, we become what we most fear. We need to balance the lack of trust that is built into relationships.

We need to learn to recognise at which point our maladaptive responses will begin to diminish our capacity for love. We need healthier and more effective ways to protect ourselves. It is about striking a balance between wariness of dangers whilst also maintaining an open, loving heart.

Sharing our lives, dreams and hearts with another person in our relationships are things that take a lifetime to master. One of the best ways to grow closer is to be truthful even when it is hard. We cannot hide the truth forever. Withholding our innermost needs and feelings from our partners, fearing we will only make things worse by speaking our hearts can take away the intimacy that connects us so dearly to those we love. Eventually, the floodgates will open and we shall discover how wrong we were.

When we can give of ourselves in loving and selfless ways, the best in us comes out. When we have not developed ourselves, challenging situations can bring out the very worst in us.

Carl Jung (Swiss psychiatrist and psychotherapist) said that each of us has a 'shadow' side – the small part of us that get scared, jealous and arrogant. The shadow is part of our nature that we are embarrassed about. We try to hide, deny, repress and avoid it, pretending it does not exist. We are all a 'work in progress.' We all have a similar dark shadow. The trick is learning how to harness the power of the shadow by becoming aware of and owning it, instead of trying to hide it. Working through it is the only hope for breaking through in a relationship.

If we allow romance to blind us to the shadow side of someone we are looking to build a future with, it will surely come back to bite us. The choice is simple. Get a handle on our shadow feelings and those of our loved ones.

When we are in love, we undo our tendency to become overly self-centred. Love creates a generous state of mind. If we have an adoring partner, he or she will help us to develop further. In a genuinely loving relationship, giving and receiving is blurred.

Our goal is to develop a more intimate, loving, trustworthy and truly generous side of ourselves. To have loved is perhaps the greatest and most honourable risk we could ever take.

Occasionally, our partners may hurt us. However, we should focus on what is left to love, even when we are hurt.

You must love in such a way that the person you love feels free.

Thich Nhat Hahn (Vietnamese Zen Buddhist monk)

4 LEARN MORE AND DO MORE

Control our emotions

Emotions affect our wellbeing. Emotions affect our body, thinking and behaviour. Positive emotions such as love and gratitude affect our health in a good way. Unwanted emotions such as anger and hatred can be burdensome, which can interfere with our lives. When we deny them or suppress them, they may eventually explode and hurt ourselves or others. We need to look deeply and identify the source of our painful emotions.

It is through emotions that our early experiences become anchored in our brains. Unless we know where these emotions and feelings come from, it is difficult to get rid of them. The key is through honest examination of our past, which will yield information that will help us.

Our emotions have a great role in determining our basic attitudes, perceptions, beliefs and thinking processes. Psychological processing of social experiences is of considerable importance for the formation of particular nerve pathways and patterns in our brains.

We sometimes engage in unkind behaviour, such as when we lose our temper; we may speak harshly even to a close friend or loved ones. Afterwards, we feel embarrassed.

Anger cannot be overcome by anger. Anger cannot solve problems. Ultimately, it will cause further difficulties. The solution to anger is compassion. Compassion will assist us to moderate our selfish and destructive emotions and thoughts.

Emotions can be strong. Sometimes, we may think that we will not survive them. Most of the time, it comes and stays with us for a while, and then goes away. Then, why should we hurt ourselves and others just because of one moment of anger? The real issue is we are not able to control our negative emotions.

We often use psychological defence mechanisms (unconscious reactive patterns) that seldom work such as denial, avoidance and blaming others. When our strong emotions come, we need to embrace it, and try to remain calm. We have to weather our own storms. The purpose is to stop and remain calm. Look deeply into the emotion and transform it. If we succeed in surviving the emotion, we can experience a more solid peace of mind. The next time the emotion arises, it becomes easier. We already know we can survive it.

Calmness comes from acceptance. If we can accept whatever is going on around us, we will be able to conserve our energy for those times we need to move into action. Calmness offers a deeper understanding such that we are at ease with ourselves and with the world. Calmness helps us to recognise and manage our emotions. It helps us manage stress better.

Calm is not being rushed. If we impulsively rearrange our lives, it often does not end well. Sometimes we have to learn to say no, step back and objectively decide what we can and cannot do. Being calm helps us be relaxed.

When we are in a calm mood, it allows the mind to function at its best. It supports rational and empathic thought. It provides the background for planning, decision-making and focused thinking. We can do our work thoroughly, make balanced judgments and learn from the past.

> *Trying to understand is like straining through muddy water. Be still and allow the mud to settle.*
> Laozi - Chinese philosopher (sixth century BCE)

Do we have the patience to wait till the mud settles and the water is clear?

Calmness offers a different path to happiness. It is about living in the present, releasing the past and letting the future unfold in its own way and at its own pace. We need to balance between choosing priorities, the present and future, or the past. Calmness represents the belief that all will work out for the best in the end, and whatever happens is for the best.

> *Sit quietly, doing nothing, spring comes, and the grass grows by itself.* — Zen saying

If we have not developed inner peace, then even if we are living the life of a hermit our minds will be overwhelmed with anger and hatred, and we will have no peace.

When we are in balance, we feel grateful and whole. We can think clearly. All areas of our brain are communicating easily and efficiently. In this state, we can take control of our lives, learn from our experiences and go with the flow. Our brain can handle stress in a way to bring it back to balance.

> *Only in quiet waters do things mirror themselves undistorted.* — Hans Margolius

Calmness is a path to inner peace.

Free ourselves from bad habits

Habits can help us and also hinder us. Habit is the best of servants, but also the worst of masters. When we engage in a habit, we are acting unconsciously, which serves to reduce the amount of conscious processing required.

We can replace bad habits with good habits. It takes mental energy to continually motivate ourselves to undertake new and positive actions such as getting up an hour earlier in the morning to exercise. However, when it becomes a habit it no longer requires much mental effort. However, our new good habits can add up to big changes in our lives because everything runs on autopilot.

Why it is so hard to break bad habits and improve ourselves?

The reason is we do not have just one brain module making our decisions. We have emotional nerve pathways that support the bad habits. Our perceptions, history and beliefs generate the inner voice from our unconscious mind that gives us excuses to not change anything.

> *It is easier to move from failure to success than from excuses to success.* John C. Maxwell

Our brain does not just follow our logical commands to drop the bad habit and replace it with a new positive action. Our emotions, histories and perceptions have significant roles in determining which response will be implemented. The unconscious mind controls our behaviour without our ever being consciously aware of what it is up to. That is why change is not that easy.

> *Habit is like an invisible thread, but every time we repeat the act, we strengthen the strand, add to it another filament, until it becomes a great cable and binds us irrevocably, thought and act.*
>
> Orison Swett Marden

How do we change our bad habits?

There is no quick step to suddenly transform ourselves. Knowing what to do and being able to do it are two different things. In order to permanently change our behaviour, we have to rewire the logical and emotional nerve pathways.

> *Knowing is not enough; we must apply.*
> J. Wolfgang von Goethe

First we have to develop a better understanding of our histories, emotions, motivations and how they interact. This is the foundation on which we can make change that ultimately uses the full potential of the human brain – a programmable structure for lifelong learning.

Second, we have to develop mental toughness to ignore the unconscious motive that gives us the excuses that it is not important to change or it can wait. The inner voice from our unconscious mind is just a different module of our brain that influences our analytical thinking and play with our mind.

We need determination to break bad habits. It is only when we take responsibility that we free ourselves from bad habits. There is no instant and definitive solution.

We are all unique and different processes work for different individuals to reprogram the nerve pathways in our brains.

We are what we continually do. Aristotle

Forgiveness is healing our past with love

We are not perfect; how can we expect others to be perfect? Everyone operates primarily out of self-interest, and that is the reason we will be hurt by others' expression of their self-interest. We need to remind ourselves that others may not always have our best interest at heart. We also need to accept the fact that we do not always act in the best interest of others.

The following story aims to provide you with the experience to see forgiveness and its benefits in action.

A man has split up with his wife. However, they had shared equity in the house they bought together. The man would like to sell the house and split the money. However, the woman was emotionally attached to the house as she was deeply involved in the design and building of the house. Therefore, she did not agree to sell and wanted to keep living in the house. The man offered the woman to buy his share so he could get his money out. However, the woman did not have enough money to buy out the house.

The impasse dragged on for a few years. During this period, the man had to contribute his half of the mortgage repayment for the house that he did not like and did not live in. When the man started another relationship, he found his financial situation to be quite burdensome. He was bitter and angry.

One day, the man sat down to meditate and came across a revelation. He realised he was not letting go and this was the cause of his distress for the last few years. He then came up with a solution, which was to accept whatever his ex-wife could pay him and give her the house. It was not logical, but the man knew it was the right thing to do for himself and his ex-wife.

He discussed his resolution with his new partner, who agreed that letting go was the right way to go. The man then contacted his ex-wife and made the legal arrangements. This action of forgiveness and letting go brought peace to all parties.

Forgiveness is freeing ourselves and healing the past.

The aim of forgiveness is to resolve past hurts and grievances. By healing the past, forgiveness can provide a more peaceful present.

How to forgive? Step by step.

The first stage to heal what has hurt us is acceptance of what has happened. It is not easy indeed. We have to accept the challenges that life sends our way. Until we have accepted what has happened, it is difficult to move forward. Actually, each hurtful situation challenges us to live as lovingly as possible. When we gain the insight that every day there are things or people to forgive, then we can practise letting go and detach ourselves from whatever is holding us back.

Forgiveness requires an open heart. We focus on how much there is still to love even when we are hurt.

The second stage to heal requires us to think clearly and wisely, to move away from adopting the victim role. We blame others' actions for our unhappiness, not our response to them.

The trauma of our childhood, youth and adulthood plays a role in our distress. However, we have a choice in how to react. It is our emotional response that makes us feel justified in our negative feelings such as hurt, anger and pain. In our minds, the other parties are responsible for how bad we feel. As we are so hurt, we are convinced it would be wrong to forgive. However, we need to understand while we cannot control the other parties; we can control our emotional reactivity. Our emotions, feelings and perceptions are not installed in a logical way. Therefore, it is difficult to remove them by rational thinking alone.

The third stage to heal is to let go. When we let go, we do not suffer anymore. This usually begins when we become concerned about our emotional and physical health. We realise that hurt and anger are not helping us and are affecting our lives. Our feelings contain the truth about our lives, and they tell us that some actions are necessary to heal ourselves.

We need a healthy body; if we do not look after our bodies, they will not serve us well. We can restore balance through taking control of our feelings and choices. Through letting go, we can remove the feeling of bitterness and the impact from others' actions or words. Then, we can gain control of our emotional and physical health. The length of time we suffer because of others' wrongdoing is up to us. We can choose to feel hurt for a shorter period of time. We can choose to use our minds to dwell in anger or to work at repairing the relationship and eventually letting go of the problematic situation. Then, we can let go of our negative feelings.

It is useful to draw upon our previous experiences and focus on the last time we forgave, whether it was last month or five years ago. Recall and relive the experience we had, in which forgiving helped us move on and feel better. Focus on how good it can feel to forgive. When we have seen the benefits and results of forgiveness in action, we are able to choose to let go of our anger over other's wrongdoing. Letting go is important because anything left incomplete can drain our energy and drags us back to the past.

The fourth stage is the most difficult yet the most powerful. We aim to become a forgiving person and make a habit of practising forgiveness with others. The aim is to learn to forgive on a daily basis and then develop a forgiving nature. It is not easy to forgive because it requires an abundance of love. Forgiveness is a step closer to loving our enemies.

> *Forgiveness is not an occasional act; it is a permanent attitude.*   Martin Luther King Jr

We achieve this when we make the decision to forgive first and let as many troubling things go as we can. We become resistant to take offence even when others provoke us. In this way, forgiveness can help to prevent future problems. When our skin becomes tougher, we shall be less likely to take personal offense. We take responsibility for our own feelings and focus on the good intentions, both our own and those of the people in our lives. It is about choosing not to be hurt in the first place, even if others neglect to do the things they promised to do. However, this does not mean we condone unkindness.

We learn to take hurtful actions not so personally, not to blame the offender for our feelings, and understand others are not perfect, and they will hurt us occasionally.

When we understand that others may hurt us, then there is no surprise when it happens. Others might be doing the best they can; when they make mistakes, the best way to help is to understand them. Forgive whatever they did that was wrong.

We do not need to wait for someone to hurt us in order to practise forgiveness. We prepare ourselves to forgive before any wrongdoing occurs. Then, when the occasion arises, we act like a forgiving person. We have many opportunities to practise forgiveness every day. For example, we forgive those who do not follow the rule and jump the queue in front of us. Our family members will upset us every day. Our relationships give us unlimited opportunities to practise forgiveness. Even in loving families, people hurt their loved ones.

Forgiveness means not withholding love or affection from them even when they do not live up to our highest standards. Forgiving does not mean we approve of everything others do, but rather that we can acknowledge that they have hurt us without making them our enemies.

Move forgiveness out of the occasional thought, and into the world of everyday action. Think like a forgiving person. Do the things we think we cannot do.

Even when we are willing to forgive, we have to apply wisdom and take care of ourselves. There are difficult individuals out there.

We all execute our forgiveness in different ways according to our own circumstances. Richard Pelzer,[60] a survivor of childhood abuse wrote his mother a letter.

> *In Christmas 1991, I decided I'm going to bury the hatchet and I'm just gonna tell Mom, "I love you as a mother but I never want to see you, I can't stand you, I don't want you to be a part of my life or my children's lives. But I forgive you. Just simply go away."*
>
> *It took days to write and I decided to mail it after the holidays. But she died on 2 January. I had it cremated with her.*

Some can live 70 years and not know how to live those 70 years. However, there are young children who have

---

[60] Richard Pelzer, author of the book – *A Brother's Journey*, which is about his survival of childhood abuse. Richard is the younger brother of David Pelzer, who was severely abused by his mother. When David was taken away by the State, Richard became the target of abuse by his mother.

suffered so much they already know how to live at a very young age.

Some of us are blessed with children in our lives. They can bring abundant wisdom into our lives. They allow us to learn and practise unconditional love. Therefore, we have to attend to them in a good way.

Both my parents died some years ago. Sometimes we understand our parents better now than we did when they were alive. When I looked back at the failings of my parents, I felt sorry for them. My forgiveness was very healing for myself. It granted me a kind of freedom. If we can understand our parents, we may capture some of the love they had for us and some of the love we had for them.

> *Children begin by loving their parents; as they grow older they judge them; sometimes they forgive them.*
>
> Oscar Wilde (The Picture of Dorian Gray)

Sometimes life is difficult; we just forgive the way life is even when it hurts.

Forgiveness is freeing ourselves and healing the past with love.

4 LEARN MORE AND DO MORE

Let go of our egos

The ego served us well in our early years. However, when we are attached to our egos and identify with our egos, we are operating from the lower levels of ourselves.

The problem is when our ego dominates, which it often tries to, it will take control over us, and we may respond in selfish ways that harm others. This is more evident in times of challenging situations, and things begin to spin out of control. For instance, in times of natural disasters such as when Hurricane Katrina hit New Orleans, social order broke down.

Our ego is like a broken compass; it can lead us to the wrong path. If a person has a very high value of oneself and regards oneself as more important and superior to other people who have different goals, ideas and beliefs, then one does not have any concern for others.

We are attached to material possessions, gains, praise, power and fame. We do not like the state of poverty, loss, being blamed and disgrace. When unfavourable situations happen to us, it makes us unhappy. But, if these unfavourable conditions happen to our enemy, it makes us happy.

Self-absorbed people tend to have unrealistic expectations about how the world will accommodate them. They are inflexible and easily angered under stress. They think they are special and the world will treat them with special favour. When they do not get what they want, it evokes resentment.

Letting go of our ego is not a one-time event. Some individuals may need a transformative event that they liberate themselves once and for all from the illusion that they are their egos.

Only when we empty our minds can we experience a life-changing revelation. Then, we can begin to expand and open to different level of consciousness. We need to discover how much a part of everything we really are.

Spiritual awareness cannot be accessed from the ego.

Selfishness is the incapacity to love anybody but oneself.

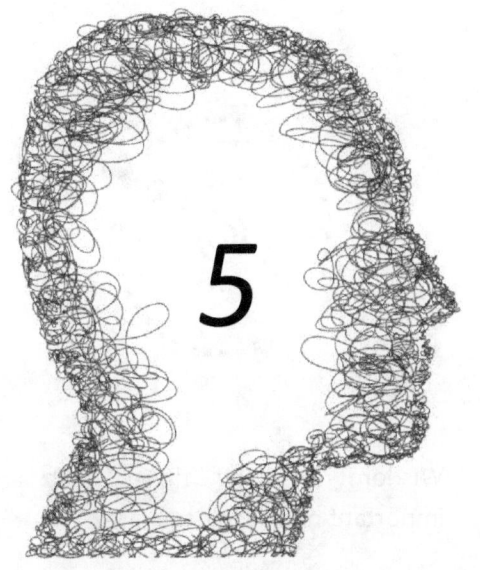

Think more

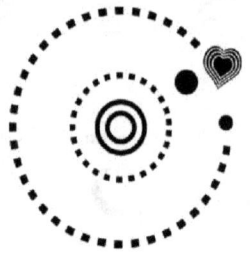

Wisdom is doing things that are important before it is too late.

Think differently

All the information coming into our brain is sorted and filtered according to our reality - our past experiences, perceptions, conditioning and interpretations in relation to our emotions, beliefs and values.

Our stored emotions are often used as a point of reference for our reactions. Our mind and emotional reactions unconsciously control our behavioural reactions. However, this may not be the best way to process incoming information.

The stored template (established nerve pathways) regarding the way we interpret the world and our associated responses may not be the most appropriate way to react. For optimal functioning, it is better for us to take charge of this process.

First, we need to better understand ourselves - how our history, psychology and philosophy interact to influence and filter all the information we receive and how we internally interpret events in our lives.

Each of us will work this through differently. Some may rely on reflection and the practice of mindfulness to identify the thoughts or mindsets that need reprogramming. Some may need gentle encouragement from others. But, some may need a harsher coach.

Second, we can consciously take control by choosing and changing the way we interpret events. This way, we can reprogram our nerve pathways. Then, we can empower ourselves with new emotional and behavioural responses to the events we experience. When we can overhaul our learned thinking, we can have different responses and therefore different results.

Our current beliefs are based on our old programming, so it is natural that our mind will rebel when we introduce a new idea to it, even though it is a positive thought. We may notice that the push back may come from our minds, emotions and bodies because the new way of thinking is getting in the way.

Though our current beliefs are deep rooted, we can correct and change them. When we exercise choice and create new intention, it creates new nerve pathways in our brains; this way we can actually change the wiring.

We can respond, act and behave as we desire with some efforts. As we understand our unconscious desires, emotions and feelings, and apply our logical reasoning and conscious thoughts to see them with an open mind, it is then possible to bring change.

Thinking differently is one of the ways to rewire our nerve pathways and connections. It enables us to see the world and respond to things that happen to us in a different way. Through this, we can rewire our nerve pathways to best support our journey through life.

When we exercise choice and create new intention, it creates new nerve pathways in our brains.

Develop insight

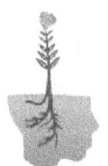

What is insight? Insight is not something that can be put into words like other branches of learning. When we acquire insight, it is like waking up to a more accurate perception of ourselves rather than holding to the former ideas or beliefs we consider true.

> *It is only after long partnership devoted to this very thing does truth flash upon the soul, like a flame kindled by a leaping spark, and once it is born there it nourishes itself thereafter.* Plato

How do we develop insight?

It seems that insights come from the same place that clairvoyance comes from. We learned that great inventors usually have a dream, or they go into a reverie and the whole answer to the problem they are working on will just fall out of the sky.

Often we expect other people to give us answers. Actually, when we look into our own minds, we will find that we knew it already. For example, is greed good or bad? We do not need others to tell us. We need to draw upon our own experiences to seek the answer.

Developing insight requires preparation and waiting for things to gel in the unconscious, which we refer to as unconscious processing. It comes from beyond our consciousness. When we are able to get out of our conditioning and have an open mind, then we will find insight.

A man told the Buddha that he considered that nothing was more important to him than his own self. Instead of lecturing this man, the Buddha stepped into the man's shoes, starting from where he was rather than where the Buddha thought he ought to be.

The Buddha suggested that if the man found that there was nothing dearer to him than himself, he should believe that everybody else felt exactly the same. The Buddha then said even a person who loves himself or herself only should not harm the self of others.

I came across a young man with a broken heart because of a failed relationship. I told him the following story with the hope that he would find his own insight.

A medical student was assigned to a mental hospital for his placement. On the first day, the supervising psychiatrist showed the student around the ward. The psychiatrist said to the student, "I shall brief you on two patients that you will be working with."

The pair walked past patient A's room. The psychiatrist said, "This patient is not sleeping, not eating, not interacting with others and is withdrawn. He is unable to function normally. He just talks to his ex-girlfriend's photo all day. Apparently, his girlfriend left him and has married another person recently."

They walked past patient B's room. The psychiatrist said, "This patient has a nervous breakdown. He married an ordinary young woman recently. Apparently, his wife drives him crazy. He thinks his life is no longer worthwhile and he harms himself."

When the pair returned to the office, the psychiatrist said to the student, "I would like to share with you a piece of information that may be helpful in the healing of patient A, if he is ready. Actually, Patient B marries the ex-girlfriend of patient A, the only woman in this world patient A thinks will make his life complete."

From the neuroscience perspective, we do not fall in love with looks alone. For instance, when a man breaks up with his first love, he will find all other women pale in comparison. He believes his ex-lover was his only true love. This is because he cannot unlearn from the nerve pathways established in relation to the pattern of attraction to his ex-lover.

We do not gain insight by accepting the opinions of other people but by finding the truth within ourselves. We can all achieve this. It is not easy, but it is not difficult either.

It is the moment when we are able to see our own limitations that we find wisdom.

Once we find the insight, we have to manifest it into the way we live and think; this is wisdom.

Don't begin to see when there is nothing left to see anymore. Baltasar Gracian

5 THINK MORE

Understand how the world works

Wisdom is to use our experiences to understand how the world works in a way that is helpful to others and ourselves.

Wisdom can help us to look at novel situations and judge how they fit into some schemes that enable us to make good choices.

The following story illustrates the importance of understanding how the world works.

A monk was walking in the hills and came upon a rattlesnake lying in the grass. The snake lunged at the monk and threatened to bite him. The monk was not afraid, and smiled at the snake. The snake was surprised by his kindness and calmed down. The monk then spoke to the snake and asked him to give up biting those who passed by.

The snake agreed to stop biting people who walked past the grassland. The next week, the monk walked by the same spot and saw the snake lying on the ground in a pool of its own blood. The snake used his remaining breath to admonish the monk, "Look what happened to me when I took your advice to be kind to others. They all tried to kill me."

The monk said to the snake, "I never told you not to hiss."

Most of our knowledge is simply rigidly held prejudice. When we do not know our ignorance, sometimes it can be the cause of suffering. Ignorance is a falsely held view that we possess the truth of reality.

> *A man must make his own arrows.*
> Native American proverb

It can be difficult to pass on our wisdom, because we may not be the person who is worth listening to. However, we can tell a story rather than telling others what to do. The listeners have to find the moral of the story and their own wisdom. They have to do it themselves.

First, they have to acknowledge their ignorance.

Second, they have to refine and modify their views in response to new information; otherwise they cannot acquire better knowledge than they had before.

Third, they have to apply the new insight to see the repetitive compulsion that causes them to make the same mistake over and over.

The solution to ignorance is through learning from others, taking in new ideas, understanding our unconscious mind, reflecting, analysing logically, broadening our understanding and deepening our insight.

It is wisdom that will help us truly change our perception of ourselves and the world.

> *We don't receive wisdom; we must discover it for ourselves after a journey that no one can take for us.* — Marcel Proust

We have to find the solutions that fit our lives.

5 THINK MORE

Nothing is perfect

We are all looking for an ideal society and a perfect life. However, most of us still get out of bed each morning and continue our lives, even though our world has little resemblance of the dreams of our youth.

We all have ideas about different societies such as capitalist, socialist or communist. However, in reality it does not matter which system we live in - the largest share of material rewards go to the most selfish. Those who work for the good of others and care for the disadvantaged are less rewarded. We search for justice, but we often see the evil and selfish prosper and the kind and gentle suffer. At times, reward and punishment is random.

Even our justice system is imperfect, and at times grossly unjust. The shocking story of Diane Fingleton[61] is one of the greatest miscarriages of justice in the Australian

---

[61] Fingleton, D 2010, *Nothing to do with Justice – The Di Fingleton story*, Chatswood, NSW Australia: New Holland.

legal system. Diane was appointed as a Magistrate in 1995 in the State of Queensland. She was then promoted to the position of Chief Magistrate in 1999. She was doing her best to try to turn the Queensland magistracy into a more accountable and efficient organisation. However, in 2002 a workplace dispute regarding the transfer of a magistrate to another town turned Diane's life upside down. The matter escalated when a Co-ordinating Magistrate was critical of Diane's leadership and her management of the transfer. Eventually Diane sent an email to the Co-ordinating Magistrate asking him to explain why he should remain in his position. The intense workplace dispute escalated into litigation, and Diane was charged as a criminal in 2002 for unlawful retaliation against a judicial officer. She was then found guilty in 2003 and was sentenced to jail for one year. Diane lodged an Appeal to the Court of Criminal Appeal (Queensland), which was refused. However, her sentence was reduced to six months, and the last six months was suspended for two years. As a result, Diane lost her position with the Queensland magistracy.

Diane spent six months behind bars, and upon her release Diane applied for special leave to appeal to the High Court (Canberra) in 2004. In June 2005, the High Court handed down the decision 6-0 in her favour, in which the judges stated that Diane should not have been investigated, charged, convicted or sentenced. Diane's conviction by the Court of Appeal of the Supreme Court of Queensland was quashed. That was based on the fact that Diane had immunity from prosecution as she was exercising her administrative function or power conferred on the Chief magistrate in the course of the work. She was just doing what she was authorised to do. It was obvious that the Courts of Queensland including the Supreme Court were ignorant of the provision. The High Court of Australia, Justice McHugh said, "It would be hard to imagine a stronger case of a miscarriage of justice."

How do we adjust our expectations in the direction of optimism without becoming naïve?

The simple answer is letting go of the idea of perfection. Sometimes, we might live in a society that favours the wealthy and neglects the less powerful. If we are unable to accept the unfairness, it generates resentment that can poison our attempts to live peacefully with one another.

Life is full of injustices; we need to find our balance between the illusion that our hopes will eventuate and the belief that life is meaningless because we cannot control it. In reality, it is more likely we do not get what we deserve in life. Also, it is likely we will not get what we expect.

When we find middle ground, together with the determination not to be defeated, we have found resilience. It does not mean we will never have a bad day; it just means that when life gets tough we know what to do to restore balance.

When bad things happen to us, we do not ask the question - why me? I met a woman whose bad fortune had been bothering her for decades. I told her to pose a different question, which was, "What do I do now in order to cope with the loss and live again?" A few years later, I met the woman and she told me she followed my advice and acquired inner peace.

Perfectionism is a distorted view, because nothing is ever good enough. When the object of our desire turns out to be disappointing, we become frustrated and unsettled. Wisdom is to understand that nothing lasts forever and nothing is perfect. In order to have a better understanding of ourselves and the world we live in, we have to match our philosophy and beliefs with reality.

Our beliefs, values and philosophy of life determine how we live, such as our responsibilities toward each other and our respect for others' right to live in ways different from our own. This determines the kind of world we create for ourselves and each other.

The pursuit of perfection can render us incapable of enjoying our lives.

Fleeting beauty of life

I was greeted by the cherry blossoms in my front yard upon returning home (Melbourne) from a visit to Hong Kong to celebrate the Mid Autumn Festival (Moon Festival) with families and friends in 2013.

For most of the year, our cherry blossom tree looks like an ordinary tree. However, in spring the flowers cover the entire tree, which is quite breathtaking. Unlike other flowering trees, the cherry blossom blooming period is very short.

The cherry blossoms reminded me of the centuries-old traditional Japanese custom of enjoying the beauty of flowers. It is the Japanese culture to turn out in large numbers at parks and temples with family and friends to hold flower-viewing parties.

The *Hanami festival* meaning 'flower-viewing,' celebrates the beauty of the cherry blossom, which mostly refers to the Sakura. It is also a time to relax and enjoy the beautiful view of the flowers blooming en masse, which look like clouds from a distance. Even just looking at a few trees, the view is intense. At a close distance, one can admire the beauty of single blossoms.

Hanami is a beautiful way to celebrate the arrival of spring. It is also a source of inspiration and a way to admire natural beauty.

The cherry blossoms are highly valued because of their transience, since they usually begin to fall within a week or two of their first appearing.

The beauty of the cherry blossoms is brief and fragile. When the wind blows, the blossoms will fall to the ground and exist only as a memory.

The physical beauty of the cherry blossoms may be fleeting but the pleasant feelings remain much longer. I visited Japan in the 1980s during the cherry blossoms viewing period, and the pleasant memory still lingers after 30 years.

At the end of the short flowering period, one may experience the complex emotions of happiness and sadness at the same time. We can learn about the fleeting nature of life from the fleeting beauty of nature.

The admiration of the cherry blossoms can be an ideal setting from which to contemplate life. It is an opportunity to seek our inner wisdom and understand the true nature of things. The symbolism of viewing cherry blossoms is to understand the ephemeral nature of life.

Once we acquire insight, we will not take anything or anyone for granted. We will appreciate that physical beauty will fade over time. However, inner beauty will stay in the heart and mind for a long time.

There are many things that are worthy of our admiration.

The most precious thing in life is its uncertainty. We should rejoice that we are alive today.

5 THINK MORE

Death, loss and grief

Most of us have a sense of how important we are to those who love us - our parents, our spouses, our siblings, our children and our close friends. When we think of the larger world, we might not do the same to strangers. This habit of averting our eyes from suffering works only until we experience our own losses and calamities. We are finally humbled when someone close to us dies. For the first time we realise that we live in a world where rewards and punishments are not allocated by merit.

When we are attached to others and they depend on our survival or vice versa, then it opens us to loss and grief; we become a hostage to our fate.

Though most of our troubles seem unique to us, they have been confronted by many others before us. Knowing this, we can respond to situations more appropriately.

Grief is intense self-absorption. After a while, we need to move on; otherwise, our lives are seriously affected. People vary widely in their reactions to grief. How long it will take to recover is unpredictable and quite individual. Some of those who have sustained great loss are defeated by it and never recover. Some will struggle back into their former selves. Some demonstrate resilience. Resilience is the capacity to respond to adversity with a determination not to be defeated by it, and do what it takes to stand up again and make a new life for oneself.

The ability to find value in suffering is a measure of resilience, which is a real edge in life. A father whose four sons died was asked about how he coped. He replied, "Life goes on."

> *Healing is a matter of time, but sometimes also a matter of opportunity.* Hippocrates

Grieving teaches us about illusion of control. We have to learn to loosen the tight reins that we keep pulling to no avail.

From the neuroscience perspective, grieving is done by calling up one memory at a time, reliving it and then letting it go. We are turning on each of the nerve pathways that were wired together to form our perception of the person, experiencing the memory, and then saying goodbye, one pathway at a time.

In grief, we learn to live without the one we love. This is difficult, because we need to unlearn the idea that the person exists.

The true test of our spiritual development is when we arise from our self-absorption and engage with the world. There are examples of those who establish memorial funds to supports victims of violence and awareness program for those who suffered. The time spent 'in the fire' can fuel our spiritual progress.

> *It is dying that makes life important.*
> Gordon Livingston

We construct elaborate belief systems to soften the finality of death and ascribe to each person a soul that outlives us. This is probably based on our fear of nonexistence after life. This longing for immortality is the most comforting and yet the most divisive way that we assert humanity.

Whatever heaven we believe awaits us, most of us do not seem to be in a hurry to get there. Instead, we cling to whatever life we have. Probably, deep down inside we have doubt about immortality as promised by our beliefs. Perhaps divinity is right here in this present moment, not somewhere else. However, religious beliefs that promise a better afterlife may help us tolerate a lot of earthly unhappiness, adversities, inequalities and atrocities.

Each separate being in the universe returns to the common source.
Laozi - Chinese philosopher (sixth century BCE)

We have to be considerate and not impose our beliefs onto others. It is possible that even well-meaning consolations such as saying to others their loved one is in a better place can be regarded as a display of ignorance. Those who do not follow our faith may find it insulting that their loved ones are part of our divine's plan to die prematurely.

I have learned something subtle about loss and love. The feelings our loved ones evoked in us shows that love is never lost. Love survives loss. Relationships don't die when the person we love leaves the planet. It is a permanent gift to us.

Try to reflect on the best qualities of our loved ones. If we can integrate those qualities into our personality, it is the best way to keep them living. When we assume the best qualities of our loved ones, they will live on.

Death is just simply saying goodbye to our existence.

死亡只不過是告別人間滋味。- 也斯

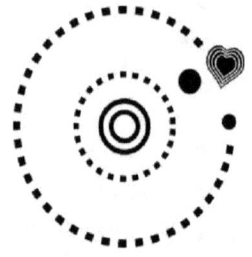

Seeing how things are too late brings no remedy, only sorrow.

Baltasar Gracian

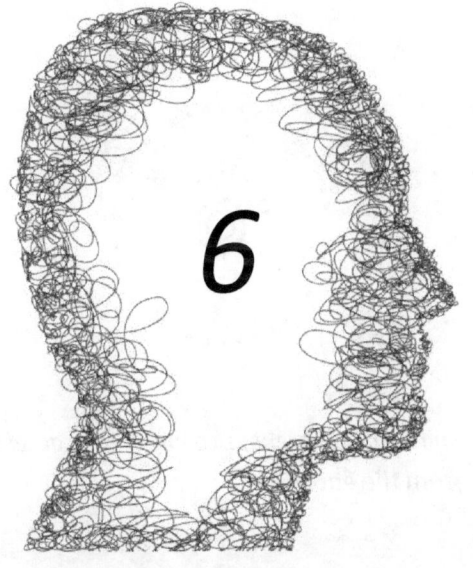

6

Become more

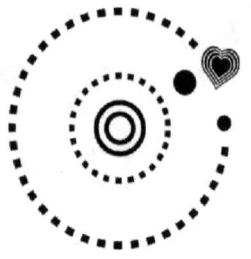

Spiritual awareness cannot be accessed from the ego.

Open our hearts

The love or lack thereof we felt as kids affects how we open our hearts as adults. If we hug the memory of our own grief to ourselves, we may close our hearts to others' misery. We may even think that our unhappy experience gives us special privileges. This is because we do not know that we can use our afflictions to make a difference in the lives of others.

The first step to open our heart is to relate to others impartially - that is people we like, people we do not like, and people we do not know. Nobody is born our friend or enemy. Friendship or enemy is a subjective feeling. Last year's friend can become next year's enemy.

The second step is to develop appreciation of the ubiquity of pain. That is to become more sensitive to the sorrow of others. We have a self-protective tendency to keep suffering at bay and insulate ourselves. We have our own ways of protecting ourselves as we navigate our way through the world. Sometimes we fail to recognise the signs of poverty, loneliness, grief, fear and desolation in others. We need to understand that besides those who are obviously in destitute situations, there are those who are suffering even while living in prosperity.

The third step is to cultivate love and compassion, and then give freely of our own resources. It requires unusual strength of the mind to generate altruism. Therefore, it is difficult to achieve the full scope of altruism.

We can learn a lot from stories, which contain the path for us to follow. Perhaps we are wired to learn from stories, in which wisdom is recorded and communicated. Stories speak to both the heart and mind. However, we have to be moved by the stories in order to learn from them.

In the book of Genesis, Abraham rushed out to greet the travellers, prostrated himself before them, brought them into his encampment, and gave them the best of what he had. Very few of us would willingly bring strangers off the street into our own homes. However, Abraham had thrown away the precautionary barriers we erect to protect ourselves from harm. This led him to a sacred experience or divine encounter.

> *Myths are public dreams, dreams are private myths.* — Joseph Campbell

Equally, those who are suffering have to be ready to open their hearts and share their suffering. They have to take the risk that others might ridicule them. It requires courage to open their hearts, expose their vulnerability and confide their most intimate pain to somebody they have never met before.

In the gospel of Luke, after Jesus' crucifixion, two of his disciples were walking together from Jerusalem to a nearby town. On the road, a stranger (fellow traveller) asked them why they were so troubled. The disciples shared the terrible story of Jesus' execution. The stranger comforted them and explained in detail the message of the prophets starting from Moses, and saying the Messiah was destined to suffer. The disciples trusted the stranger to enter their hearts. When they arrived at their destination, the disciples offered their new friend to spend the night with them. The moment of recognition came when the disciples shared the meal with the stranger; they understood that all the time they had been in the presence of the Messiah.

Wisdom is often recorded in stories, but we have to find it ourselves. We will recognise its truth only when we put it into practice in our own lives.

The above stories reminded me of the moment I reached out to embrace the pain of another person who felt unloved. As Mother Teresa said, "The most terrible poverty is loneliness, and the feeling of being unloved."

At the end, I found that when I opened my heart to others, that enabled me to transcend the overwhelming memories of my younger years through alleviating the pain of others.

A few years later, I met that person again. She had a smile on her face. It was a smile of a person who had accepted her life and cherished herself in a new way. I was very happy for her. What could be better than the feeling of being loved?

We can try to make friends with a distressed person, so he or she becomes a presence in our lives. We can make a significant difference for others. Our involvement in others' suffering and alleviating the distress of others will also make us happier.

On one occasion I delivered a seminar to a community group in Hong Kong. After the seminar, one of the audience members came forward and shared her story with me. The young woman told me about her predicaments. She told me her father was the cause of her suffering and there was nothing she could do about it in the foreseeable future. She was filled with tears. She felt no one could understand her.

I tried to put myself in her place and told her sometimes it could be a long wait before we get out of difficult situations. I knew that because I have waited for decades for my own suffering to pass. It was painful because I did not know how long I had to wait. The woman sensed I could feel her pain and felt deeply understood.

It can be difficult to ask for help. It takes courage to ask for help. Some of us do not like to open our hearts because it may put us in a vulnerable position. Our pride may make it difficult for us to receive kindness from others. However, there are situations in which we need others' assistance. We need to know when to receive as well as when to give.

Opening our hearts requires us to leave our egotistic selves behind and overcome our inertia, reluctance or fear. For those driven by excessive fear of making mistakes, every time they are invited to embark on a relationship, they consciously or unconsciously ask themselves whether it is safe to show the other person who they truly are. They ask themselves whether they should entrust their hearts to the other person. They keep asking themselves because they do not want to put their hearts in the care of another person who may not be genuine.

The ability to recognise important feelings in other people, such as joy, fear, disgust, sadness and pain is genetically configured in our brains. However, this gift is not equally well developed in all of us.

There are those who lack the ability to put themselves in the place of other people and empathise with others' feelings. Their feelings, thinking and behaviour are primarily determined by self-centred considerations.

We have to be willing to and also possess the sensitivity to place ourselves within another person's world of feelings. It is this capacity that sets the human brain apart from all others.

To open our hearts, our aim is to develop the ability to enter through our inner feelings and that of another person. Ideally, we need to develop the ability to enter into the feelings of many others.

When we forget about ourselves, our hearts will be empty of self-importance. Without the distorting lens of selfishness, this naturally leads to empathy. Empathy requires a tremendously refined level of perceiving and processing other people's nonverbal expressed feelings.

> *The perfect man has no self, once he has lost the belief that he is special, he regards all other people as "I". People cry, so he cries – he considers everything as his own being.*
>
> Zhuangzi – Chinese philosopher (369-286 BCE)

In order to function at a higher level, sometimes we have to take a leap of faith; sometime we have to let go. Both require us to open our hearts and minds, and to embrace new possibilities.

Our hearts or kindness for others have no limits; the more we give, the more love is created.

Opening our hearts to others are moments of spiritual insight. We shall be enriched by the encounter.

6 BECOME MORE

Open our minds

Narrow-mindedness is the worldly way of life. However, we have to realise that our beliefs are just thoughts rooted in the past, which may not be true or correct. To be open-minded is an experience beyond our beliefs.

When we are open-minded, we are more open to communication and able to pick up messages and meaning from other people around us. When we open our minds, we can move to a higher level of consciousness.

I once had a conversation with an elderly mother, who spoke unkindly of her adult daughter. They had not seen each other for a few decades. I did not have adequate knowledge about their family dynamics or events that happened in their family. Therefore, I listened with an open mind.

Years passed and through a convoluted process and an exceptional brush of luck, I met the woman's adult daughter. The woman's daughter shared with me accounts of her stories, which were very different from her mother's version of events. She did not have any bad feelings towards her mother, who did not treat her well. I chose to believe her version of the story. That was based on the principle that if a mother genuinely loves her children, why would she speak unkindly about them?

We can learn to suspend our judgment and truly listen to the merits of what others are saying in an unbiased way. It is like the instructions a judge gives a jury.

> *It is important that you keep an open mind throughout this trial. Do not form or express an opinion about this case while the trial is going on. You must not decide on the verdict until after you have heard all the evidence and have discussed thoroughly with your fellow jurors in your deliberations.*
>
> *You must decide what the facts are in this case. Your verdict must be based only on the evidence that you hear or see in this courtroom. Do not let bias, sympathy, prejudice or public opinion influence your verdict.*

We are entitled to our own opinions and emotions. However, when others are talking we are tempted to react. Actually, what is required for good listening is to temporarily put our own feelings and opinions aside in service to the other person.

There were a few occasions that I have met people who were exceptionally certain that their views were the only truth, even in the absence of evidence-based knowledge. Often, their beliefs did not come with modesty. These individuals hold unshakeable beliefs that they are absolutely correct - their religion, political and philosophical views are superior without consideration of others and the reality.

Conversations with these individuals often turn into unpleasant discussions and derisive arguments because their sole purpose is to indoctrinate others with their views and beliefs. They do not understand aggressive debate is unlikely to change hearts and minds.

We do not need to control everything; relinquishing control is one of the greatest gifts we can give ourselves and those around us. Control is actually an illusion.

People who are short-sighted and superficial in their thinking and narrow in their perceptions are not functioning at a higher level. They are making limited use of the gift of the human brain. They suppress any concerns regarding their mistakes, and turn away all doubts concerning their thinking and behaviour. The ways they behave have actually taken away their capacity to learn and grow. If we do not want to follow their path, we need to honestly ask ourselves the following questions.

Are we ready to change our views if the evidence is sufficiently compelling?

What is our goal? Is it to win the argument or to seek the truth?

What is our intention? Is it to defeat and humiliate the other party with malice?

Are we listening in order to twist others' words or use them to further our own cause?

Are we making place for the others in our minds in the Socratic manner?

The preferred mode of exchange of ideas is the Socratic dialogue. This is designed to produce a profound change in the participants, in terms of gaining new insights. The purpose is that everybody would understand the depth of his ignorance. It is a spiritual exercise, which is conducted in a kindly and compassionate manner. Questions and answers are exchanged in good faith and without malice. Therefore, it is not about winning. The manner in which thoughts are exchanged allows others to unsettle their own convictions, and make a place for others in our minds. It is about listening intently and sympathetically to the ideas of others. It is not about what we feel, what we may think or whether we agree. It is about acquiring a true understanding of others' positions. This way, we shall be open to wisdom no matter how it arrives. We will not discount inputs coming from other people or sources outside our own small circles or rigidly held beliefs.

The treatment given to Galileo by the Catholic Church in the 17th century illustrates that close-mindedness can breed fear, hostility, intolerance and even violence. Galileo conducted experiments to prove his theory that the Earth revolved around the Sun. He knew he was correct and the findings were true, so he published his theory. However, the Catholic Church believed otherwise and regarded him as a heretic. The Catholic Church was very powerful at that time, and Galileo was forced to confess he made a mistake, which he did because he wanted to save his life.

Openness is the path to peace, humility, and understanding. Try listening to someone whose belief system we do not agree with and find out the similarities to ours. Go to a religious service that is different from ours, and find things there that we can consider holy.

In December 2013, I travelled to the Mount Gambier region in South Australia to have the geological experience of the volcanic region. At the Tourist Information Centre, I learned about the Australian Aboriginal legend regarding the volcanoes through the works of Christina Smith,[62] who documented the lives, legends and language of indigenous Australians from the Mount Gambier region.

The Aboriginal Dreamtime (spiritual) legend talks about the giant ancestor of the Booandik Tribe, who long ago camped and made ovens at Mount Muirhead and Mount Schank to cook for his wife and family. However, the evil spirit at each site frightened them. So they moved to Mount Gambier, where they made ovens and lived a long time. One day, water rose and the fire went out. The tribe then dug other ovens, but each time, water rose from below and put out the fires. This occurred four times and created the present Valley Lake, Blue Lake, Browne's Lake and Leg of Mutton Lake. Finally, the giant and his family settled in a cave somewhere nearby.

It may be difficult to accurately date this Aboriginal legend, but the story implies volcanic activity took place comparatively recently in geological terms. It is estimated volcanic eruptions occurred between 2,000 and 7,000 (BCE) in that region, which coincided with the Aboriginal legend telling of ovens. Scientists estimated

---

[62] Christina Smith (1809-1893) was a teacher and Christian missionary. Her book, *'The Booandik Tribe of South Australian Aborigines: A Sketch of Their Habits, Customs, Legends, and Language'*, was published in 1880.

around 1,800 (BCE) volcanic explosions formed large craters at Mount Gambier, and subsequently turned into lakes.[63] This finding matched well with the Aboriginal legend that the ovens were swamped by water.

The stories of Australian Aboriginals are passed down through generations orally; they were never written. However, this Aboriginal legend seems to link well with the geological events about the volcanic activities in that region many years ago. It shows that at least this Australian Aboriginal legend may be more than just myth.

When we see the limitations of human perception and thought, we shall be amused by the manifest of narrow-mindedness.

With an open mind, we can keep the mirror of self-knowledge clear so it can clearly reflect the errors we make.

Find merits in what others think, feel or do.

---

[63] Office of Minerals and Energy Resources South Australia, 2001, *Volcanoes of the Mount Gambier Area. Earth Resources Information Sheet July.* South Australia.

6 BECOME MORE

Hold onto our integrity

I once received an e-mail from a woman to whom I had offered support. It was a harsh message, but I did not understand fully the reasons behind it at that time.

Two years later, I received a phone call from the woman and she told me that she had broken up with her partner. She was in a miserable situation and she sought support from me. She also apologised for the e-mail message sent two years ago. She told me her ex-partner manipulated her and he sent the message disguised as her. The woman acknowledged she had acted unwisely.

I told the woman the e-mail message did not matter at all, because once the act of giving is done we have to detach from it and move on to the next moment. Giving in any other way is not pure.

When we are able to show others how we have expressed what we believe, this is the best way for others to learn about our integrity.

There was a long period of time in which my reputation took hits through vicious rumours and damaging gossip within my family. My sister and brother are the only persons who have trust and faith in me. They would not listen to the lies, slander, falsehoods or allegations that others perpetuate.

Over a number of years, I made regular visits to one of my elderly relatives who lived overseas. One day, she burst into tears, and said to me, "I am very sorry. Please forgive me." I did not understand what she was referring to. She then said, "I have been foolish throughout all these years believing in what others had said about you."

Reputation and integrity are closely connected, but quite different. At the end of the day, our reputation may be bruised. However, our integrity is harder to attack regardless of what others say. Our integrity and trustworthiness to others will remain intact if it is established and strengthened over the years.

What I have learned is that integrity is power. Integrity dignifies our existence. It is more precious than bravery or great intelligence. To claim any virtue, we must manifest it in our behaviour and in the way we live. We should not allow the enemy of the moment to frighten us into actions that violate our core values, or cause us to lose faith in ourselves. We have to earn our trust and respect over time.

We need to develop ourselves continuously because if our integrity is not strongly established, it may not hold under difficult situations. Usually, when a crisis happens there is little time to respond wisely. When we are guided by our integrity and inner wisdom, then we will not be disorganised during the crisis as we already know the most appropriate way to act. A common example would be public service officials giving into temptations to solicit or accept bribes, breach public trust, or dishonestly and improperly use knowledge, power or resources for personal gain.

When we have developed ourselves, we would be able to say no to temptations right away rather than thinking whether we should take the bribe or how to cover it up so no one will find out. The only sure way no one will know about our dishonesty is if we never commit any deed that is dishonourable.

Values are what we believe is important, and they underpin all of our lives. A life well lived requires integrity, which is consistently applying values that are important to us. Our true selves are expressed through maintaining our integrity. Integrity is the heart of who we are.

If we lose our integrity, we have nothing more to lose.

Connect to our calling

We need to ask ourselves what really matters to us and who we really are?

Connecting with our calling is one of the best ways to feed our spirits. The burning desire to be or do something gives us the energy to get up every morning. It also enables us to pick ourselves up and start again after disappointment.

A calling is a mission, a special assignment given to a person especially suited to perform, which at the same time affects the larger world in a positive way. Calling is meeting our greatest joy while simultaneously meeting a genuine need out in the world.

> *Great gladness meets the world's great need.*
> Frederick Buechner

Those who found their calling would tell us something like this, "I am created to practise in the nursing profession to help the sick." Those who work in caring occupations, in which they help and serve others, carry the belief that their efforts have improved the lives of the people they encountered. A calling can be as simple as to be a caring parent. It can also be an involvement in community work to help children who are homeless.

> *Every calling is great when greatly pursued.*
> Oliver Wendell Holmes

We need to find something worth living for, something that keeps us moving ahead no matter what obstacles we face. Finding our calling is one of the critical milestones in our spiritual journey.

Kahlil Gilbran wrote, "Work is love made visible." He referred to calling as integrating work and personal fulfilment. It is about finding a way to do what makes us happy, making a living and making the world a better place at the same time.

In reality, a calling doesn't always lead to a job. However, we all need a paid job so that we get money, pay our bills and survive in the modern world. Few of us are lucky enough to have work that provides real satisfaction.

Most people do not derive a lot of psychological satisfaction from their work. They just see work as a primarily routine activity that feeds their families and allows them to pursue more enjoyable activities in their free time.

Since 2010, I have devoted time to share my experience with others and make a contribution to relieve others' suffering, which I regarded as my calling. It is not paid work at all. However, the rewards I received were great - they came from learning others found comfort and benefit from my effort.

Our goal is to find a passion bigger than ourselves to dedicate our lives to.

> *We make a living by what we get, but we make a life by what we give.* — Winston Churchill

Our humblest hour is when we compare our life as it is with what we vowed to make it.

Grow our souls

Most of us do not focus on the question regarding the importance of meaning in our lives. Spiritual or personal growth does not just happen. By making the effort, we can come a little closer to answering for ourselves the fundamental question of human existence - why after all, are we here?

Why should we take the trouble to embark on the difficult path to focus in our outer and inner worlds as accurately as possible?

The happiness and suffering each of us experiences is a reflection of the level of distortion or clarity with which we view ourselves and the world.

Any desire to pursue a spiritual path of inner transformation will arise only when we recognise the underlying miserable state we are in. To change our lives, we must first acknowledge that our present situation is not satisfactory. We need to understand ourselves, and that our behaviours propel us to act in ways that cause more suffering. We have to learn from our mistakes.

What is required to encourage unselfish behaviour?

All the stories of great men and women led to the same moral. They all accepted the invitation to endure all sorts of challenges. The sufferings they experienced in their real world journeys fuelled their progress on their inner journeys. For example, Nelson Mandela's outer life journey of 27 years in prison was the foundation for the flourishing of his soul.

Kindness and courage are virtues that are not normally distributed in the population. Some people gave their lives for the idea of equality or democracy, while others just stayed home to watch a movie. Generally, our own real-life stories are less dramatic.

We are like a piece of clay, even though it has been shaped and painted with glaze it is not very useful until it is put in the kiln and subjected to intense fires. It is the time in the fire that makes it strong.

> *What is to give light must endure the burning.*
> Eleanor Roosevelt

It is suffering that transforms us and rewards us with wisdom. Suffering is the vehicle for spiritual growth. Deep down, we know it is our path, too.

We have to learn the important lessons the hard way; it is through enduring that we get past the suffering. It can feel unbearable as though we cannot take it anymore and that it will never end. It is natural to want to escape from it any way we can, but the only real way out is to go through it. It is wise to use the time in the fire to grieve as intensely as we can. The pain is the fuel for our spiritual growth. When we look back, the hits we took in the end taught us what we need to know. When we see the meaning of the hits of the past, then we need to apply the wisdom to see the same meaning in the present.

How does true love grow?

There are lots of people who are suffering. It is in the newspaper every day, but some of us do not feel a sense of deep concern about it.

True love grows from developing a deep concern for others, and the willingness to bear the burden and sacrifice to help others relieve their suffering. It is through cultivating the mind to devote our lives to others different from ourselves.

Compassion is the face of altruism. It is a feeling from deep in the heart in which we find we cannot bear others' suffering without acting to relieve it. We all benefit from genuine expressions of selfless love. We intuitively know this to be true, and there is scientific proof supporting this as well.

How can we help others?

We can help others by providing food and shelter. We can help others by teaching them new skills. We can help others by opening our hearts, showing compassion and removing others' suffering.

How do we prepare the ground work?

Like a Christmas present, most of the time we work on the wrapping, but if what is inside the box in empty it is not of any value. Like ageing, the shiny wrapping will gradually fade and the only beauty left to show is from the inside. In the end, the external emanates from the internal.

We need to develop the capacity to put ourselves in the place of other people and to empathise with them. If we want to be a friend of all people, then generate love and compassion. Compassion is complete if applied to all beings with no conditions. We must change ourselves and practice love and compassion. If the internal is full of love and warmth, then external problems can be accepted and more easily faced.

We all want to be special and to have a dignified existence. However, this requires us to put in efforts in order to leave behind a footprint to mark our passage upon the earth.

Wise people think of others, foolish people think of themselves.

The story of Alfred Deakin (1856-1919), Australia's former prime minister,[64] can inspire us to learn to grow our souls. He introduced the idea of a minimum wage that recognised workers as human beings and equal citizens rather than treating them as commodities or mere units in the cost of production.

When Alfred Deakin was a member of the Victorian Legislative Assembly in 1896, he spoke in support of the introduction of a minimum wage. That was at the end of the 19th century, in which workers were treated unfairly in all parts of the world. Some had been slaves, or treated as slaves, coerced into contracts that denied or restricted their freedom, or forced into underpaid labour. Alfred Deakin said that it was not only a matter of social justice, but essential to our equal dignity and mutual respect as Australian citizens. He said, "We were not joined simply by economic transactions. Citizenship entailed a duty of care and relations of reciprocity and mutual obligation. It would demean us all, if those who made our food and clothing or tended to our comforts and wellbeing were treated as inferior beings, unworthy of our care."

---

[64] First term 1903-04, second term 1905-08, third term 1909-10.

However, selfishness is still striving in each generation. Professor Peter Brooks[65] lashed out at his overcharging colleagues (medical specialists), who charged exorbitant fees six times higher than the health funds rebate or schedule fee under the Australia Medicare Scheme. Professor Brooks wrote, "These exorbitant fees are levied on the most vulnerable in our society, those who are unwell and require medical attention." He further said, "Having taught medical students for 50 years and being a specialist, I am embarrassed that my colleagues behave the way they do."

It seems there are some of us who think they are too intelligent to be kind. Our source of morality could be philosophical, religious or situational. However, we all know that protecting the weak from exploitation by the strong is a fundamental obligation of any civilised person.

Why should we try to know ourselves and ultimately to become conscious of what is taking place in ourselves?

When we are conscious of who we are and how we have become what we are, it will enable us to take control of

---

[65] Brooks, P 2014, *Submission 13 - The Senate Select Committee on Out-of-pocket costs in Australian healthcare*, Parliament of Australia, Canberra.

our own lives. Then we shall be our true selves. However, our highest self is not the end. Our development never ends; there is always growth, and another goal to accomplish.

How do we know we have grown?

As compassion grows stronger so does the willingness to commit ourselves to the welfare of all other beings, even if we have to do it alone and make sacrifices. Our growth in patience, tolerance and love for others shows how far we each have come as a spiritual being. Also, as we are changing we gradually attract different friends. Our old friends who once shared the values and interests we had seemed to be diverging. Perhaps, we may also change our jobs as well.

> *The greatest reward for a man's toil is not what he gets for it but what he becomes of it.*
>
> John Ruskin

If we have not made an effort to grow our souls, at the end of our lives we will face the unpleasant and uncomfortable realisation that we have not made full use of ourselves throughout our existence.

We do not need to be a very intelligent person - what we need to be is a kind and loving person; that is a wise person.

6 BECOME MORE

Become a bit kinder

What does it mean to live a small life? A small life is one that emphasises material success.

What constitutes greatness in our lives?

It is kindness. With kindness we shall live a life that benefits others, which is based on caring, giving and the positive influence we have on others.

Yu Qiuyu wrote, "Finding the true meaning of life requires a special vision and kindness." [66]

Our existence can be of value to others. This gives meaning to each day of our lives, infusing us with gratitude and readiness to give. The way we spend our days reflects how we spend our lives. Infants, children

---

[66] Yu Qiuyu (余秋雨), contemporary Chinese cultural and literary scholar.「人人都在人生中，但發現人生，卻需要特殊的眼光，甚至，特殊的仁慈。」

and adolescents are dependent on the kindness of their parents or caregivers. When we age, we depend on the kindness of others. However, in adulthood we may provide care to others, but our friends, family members and the community provide us with other things we need. We are interdependent. There are thousands of people in our lifetime we may never meet who are showing us kindness. We may falsely believe we are independent, but this is not true.

I was exceptionally moved by the scene in the movie, *Les Miserables*, in which the main character, Jean Valjean, was imprisoned for years for stealing bread for his starving family. Upon release, Jean was unable to find shelter because of his status as a former convict. Eventually, Jean knocked on the door of the church and the Bishop offered him shelter. However, in the middle of the night Jean ran off with the church's silverware. Subsequently, the police captured Jean and returned the stolen goods to the church. However, the Bishop told the police that the silverware was given to Jean, and that he had forgotten to take the other silverware as well. The police accepted the explanation and released Jean. The Bishop then told Jean that God had spared his life and he should make good use of the silverware to make an honest man of himself. Jean's life was turned around by the Bishop's kindness. He then assumed a new identity and pursued an honest life, giving of himself for others.

In my travels, sometimes I was approached by strangers for assistance. Though I was warned of people pretending to be in desperate situations, the scene from the movie, *Les Miserables* inspired me to think again before turning away anyone who asked for help.

I value the time I listen to people's stories, engaging with their emotions, and understanding their stories. Occasionally, I question their fundamental beliefs, and try to help them identify the real cause of their suffering and develop insights to shape their own lives in a more positive way. It is fulfilling to know there is one less soul that is suffering.

The universal philosophy and wisdom is quite simple - it is kindness. There is no need for complicated philosophies, teachings, dogma and beliefs, temples, churches and other places of worship. As the 14th Dalai Lama said, our mind and heart is the temple. Without appreciation of kindness, society breaks down.

How do we develop kindness for others?

It seems that young children can do this better than adults. They can love unreservedly and selflessly.

We have to be mindful of our tendency to create categories in people we meet. When we group them as friend or enemy, this restricts love. Our enemies can provide us with opportunities to practise patience, tolerance and forbearance. We also tend to divide people into those who are more attractive and those who are not so, and show more compassion for some but less for others.

First, we start with becoming a friend to all people. Have a concern for their situations and be ready to offer assistance if required. This is through cultivating the concern like a dedicated mother feels for her child; then direct it towards more and more people. When we enhance the sense of closeness with others, we can increase the intensity and finally extend our love.

Second, we learn to love by acknowledging how others suffer and then develop compassion, the desire to see others relieved from suffering. It is the strong wish to see others happy. In order to expand our love for others, we have to develop compassion toward those for whom our feelings are neutral, non-existent, weak or even negative. Love creates a feeling of connectedness and solidarity. It is a feeling that keeps spreading outward until in the end it includes everything that brought us into the world. The aim is to forsake self-centeredness.

Third, think about the kindness others have given us, we shall then reciprocate and extend it to others. Hearing stories of great kindness may be the first step. It can establish the powerful predispositions of compassion.

To open ourselves to other people and new experiences requires a special sort of humility that resists the forces of our egos and limited vision. We all need other people with whom we can share our feelings, experiences, knowledge, insights, joys and sorrows. This way, we live our lives well. However, this requires us to learn to love.

We can try to take one thing that touched us to act on it. It could be one simple act, or it could be a great act as part of a large project with others. Choose a path that is appropriate to our abilities, and try to offer the highest level or degree of love we can. Our aim is to help others.

Love makes us feel connected with all things and grateful for what nature has given us. We are here to grow in love. It is never too late to say "I love you or I forgive you."

We will never be free from the chaos of life, just like the clouds will always move across the sky unpredictably and beyond our control. The only real freedom we shall have is to love more freely.

Very few of us leave anything behind that is memorable to others. It would be good enough to have loved those we could, done as little harm as possible, and given hope to others.

There is no happiness without kindness. What we do for others, we ultimately do for ourselves.

What we need is humility

Wisdom is acquired through focusing on the big picture rather than our traditionally held beliefs. We need to rise above our personal needs and show compassion for others. We need to understand the importance of interdependence with others and the collective good. Letting go of our ego, showing kindness, forgiveness and humility are part of wisdom. From the neuroscience perspective, this requires us to recruit new nerve connections and rewire our brains. We have to keep doing it until it becomes effortless. This is the way to make our lives richer.

There is a wide spectrum of beliefs in the world's theologies. There are many great things that have happened in the name of faith-based organisations. There are equally many evil things that have happened in the name of faith-based organisations that they do not acknowledge, address or recompense.

Some belief systems are rooted in a profound fear of annihilation. Belief systems that claim to be the sole and exclusive owner of the truth form the kind of self-righteous and superior approach to life that not only divides people around the world, but also divides us internally. It drives a wedge between us and within us. When a rigidly held belief is claimed as truth without humility, we are then polarised into groups. It is unholy to say, "This is the truth about God and only our belief possesses it." Actually, the holy approach to faith does not slam doors in people's faces.

Fundamentalists from any creed believe they hold the copyright to the truth and they must impose their truth on others. It is arrogance more than humility. They look at others with harsh judgment. They show arrogance even within their belief system. Each side believes that they are right; they are the sole possessors of the truth. Each side interprets events in a way that serves their beliefs. Each side has answers to everything.

We can hold the unknowns gently and allow them to unfold. We can clench them with a closed, tight fist, trying frantically to control everything and driving ourselves and others around us crazy in the process.

Those who are not open-minded will group with like-minded people, because it is easier for their brains to handle. The potential trouble is when these like-minded people add up to a very large number, the other 'inferior' people might be declared to be their common foe and are ostracised.

Eventually, this large group of people reinforce each other in such as way that they do not have any concern or doubt about their correctness. They cannot feel that their actions are hurting others. That is contrary to the belief that teaches them to love one another.

Some of us worship their god and follow the rules to attain everlasting life. Then, they declare the gods worshipped by those unlike theirs as not true. Sometimes, the intolerance may grow out of control and may even be followed by violence.

We learn about the world from people with whom we agree politically and philosophically. Unfortunately, some of us are accidental victims of childhood indoctrination. The environment in which we grow up is more or less consciously structured in accordance with the standard of previous generations. We are born to particular parents, in a particular place and time and raised to believe in our parents' god to the exclusion of all others. We may see that others as just wrong and we may be eager to impose our beliefs on others. However, very few are able to see this.

Jean-Jacques Rousseau opined that because we could change we did not always know what was natural in us and what was acquired from our culture. We could be overly shaped by culture to the point where we drifted too far from our true nature.

There are many paths to the divine. We are allowed to custom-design our faith to fit the circumstances of our lives. It is our own sacredness. Our journey is not another's journey. It seems straightforward, but it requires a humble and gracious heart to resist imposing our own standards onto another.

Actually, there is a common virtue in almost every faith - humility. However, not everyone is aware of this. When people hold their beliefs with humility then different beliefs stop being a dividing factor. On the contrary, beliefs become a gift.

Religion is a source of wisdom, not an excuse to spread judgment and hatred, to exert power or to impose our version of truth on others. Our faith allows us to develop wisdom, to help us to grow and deepen our understanding of ourselves and others. We need to learn to graciously and humbly invite others into the wealth of our faith and share the wisdom. One way to develop humility is to practise saying, 'My thinking is partial and incomplete. I am reflecting on the part of the truth that I have come to understand, but I am open to continued discovery. If you choose a different path, I will do my best to understand and respect it.'

No religion, science, spiritual, or philosophical system is complete; the mystery is still unfolding. Wisdom and perhaps faith can be used to wrap our hearts and minds around all of life's unknowns.

We have to ask ourselves, how tightly do we hold onto our faith as superior to that of others?

The 14th Dalai Lama once said if scientific discoveries conflict with Buddhist doctrines, the doctrines must evolve with these discoveries.

I have the fortunate opportunity to meet with an old friend who selflessness I admired. She was involved in charity work with a religious organisation for years. I put forward my view that the members of her organisation actually would try to convert those who helped by them to believe in their God, and if that was accurate then their act of love or giving was not unconditional. She humbly replied, "I have been reflecting on this for a while."

Whether or not we believe in a previous life and an afterlife, we need to live a moral and ethical life. All religions seek to find compassion and wisdom. Our greatest challenge is to awaken and search for the mystical sense of union with others. Our goal is to let go of our cherished beliefs that we know are absolutely true. We need to loosen up our attachments to our beliefs and see what happens.

We can be humble and yet at the same time not compromising what we have chosen to believe in.

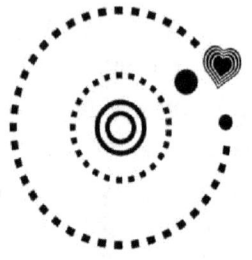

One step out of our way and the way appears.

Laozi - Chinese Philosopher (sixth century BCE)

Author's note

I realise I only know a little about life. However, I have not waited for my life to be perfect before I share what I know. This will take a lifetime and I shall never be perfect even by then. We are all a 'work in progress.' Therefore, I try to do what I can, sharing with you based on the best knowledge I have and the life experiences that shaped me.

This book brings my own experiences and philosophy of life, in addition to findings of neurobiology, neuroscience, psychology and psychotherapy as a way to gain better understanding about ourselves.

I met people who were suffering. I met people who cried every night until there were no more tears. I met people who did not show any tears regarding their suffering. When I asked why they did not have tears, they replied, "When there is nobody to hug you when you cry, eventually you stop crying." I have also met people who were unable to tell the stories about their sorrows.

Writing serves my urge to communicate and make sense about lives that are confused, lives in the shadows, lives that are difficult or do not make sense. Therefore, I started writing books. It is my contribution and responsibility to others. I read a lot and I felt that books are of great benefit. The wisdom of great men and women who have come before me reach out through the books to support me as I read their words. As time goes by, I realised that I wrote the books for myself as well. We teach what we most need to learn.

I was able to establish close connections with others who were suffering, because they felt I fully understood them. There were few occasions I told others they might have to wait patiently for a very long time, sometimes a few decades for difficult situations to go away. However, none of them challenged what I said. Eventually, I asked one person why she accepted what I told her. She said, "You are speaking from your heart. I can feel it."

After I embarked on my spiritual journey, I have less commonality with people who continue their search for material success, fame and power. However, I meet new friends who are on their spiritual journey. It is uplifting to meet them. We do not need a few thousand friends or a million 'Likes' on social media. What we need is just a few intense and meaningful encounters with other individuals that will make it possible for us to increase our bank of experience and grow our souls.

I hope this book will get your heart and mind in harmony, on a path toward inner peace.

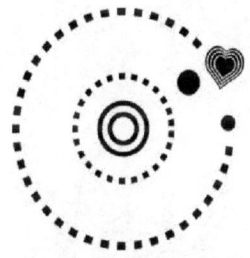

Every day there are things we should be grateful for.

Every day there are things we need to forgive.

Other Titles

Paperback: 208 pages   e-book also available
Published: September 19, 2011
Language: English
ISBN-10: 1466229810
ISBN-13: 978-1466229815
Product Dimensions: 8 x 5.25 x 0.5 inches

For more information and sample, please visit
http://www.franciskung.net

Ways to be more loving

Kung, Francis (2011). **Ways to be more loving**. CreateSpace, South Carolina, USA.

This book provides the signposts and practical wisdom for us to develop ourselves to be a more loving person. Love is beyond romance, when we develop wisdom, we would understand that our goal is to love others, not to find someone to fill up our need for love.

This is a valuable resource that would enable us to enrich our relationships. It is written for young adults, parents and people at different stages of a relationship, those who feel alone or lost their directions and those who would like to develop themselves to nurture others.

"It is love that motivates me to write this book." Dr Kung

What others have said:

"You are writing from your heart. I can feel it."

"I am beginning to understand what love is."

"This book can bring love and hope to the world."

Paperback: 132 pages   e-book also available
Published: February 29, 2012
Language: English
ISBN-10: 1468153595
ISBN-13: 978-1468153590
Product Dimensions: 8 x 5.25 x 0.3 inches

For more information and sample, please visit
http://www.franciskung.net

Ways to better living

Kung, Francis (2012). **Ways to better living**. Second edition. CreateSpace, South Carolina, USA.

This book represents a universal philosophy of the purpose of life, the way life works and ways to better living. Citing a range of well known teachers, philosophers and leaders from across the ages, Dr Kung weaves his own wisdom throughout in an easy to read and enjoyable way.

What others have said:

"I like your unencumbered approach to live a better life."

"It touches the deep side of my heart. It is a treasure for me."

"I am so glad that you have put down how you feel and what you believe in such a clear, concise and knowledgeable manner to share with others. I am sure that a lot of people will find it comforting and inspiring. It is to me."

Paperback: 248 pages   e-book also available
Published: March 18, 2013
Language: English
ISBN-10: 1482350395
ISBN-13: 978-1482350395
Product Dimensions: 8 x 5.25 x 0.6 inches

Love Work Wisdom

Kung, Francis (2013). **Love Work Wisdom**. CreateSpace, South Carolina, USA.

The purpose of this book is to make a contribution to others' happiness and wellbeing. This book is for anyone who aspires to let love and wisdom to fill the gaps and guide his or her life. Our task is to open our mind and search for balance between polarised views.

Then, we shall be able to see the big picture and appreciate the secret of life. We shall also find our directions to the big questions. What is important in our lives? What should we do with our lives? What meaning will we make of our lives? Our inner wisdom can help us overcome our blind spots, widen our horizons, and develop balanced solutions to our problems. In the end, we will find answers to our day-to-day questions. Why am I not happy most of the time? Why am I angry with myself, my partner and my family? Why do certain situations make me nervous? Why am I not satisfied with my job? Why do I keep failing with relationships?

When we follow the path to a higher level of being, the journey will soften us inwards and at the same time expand us outwards.

You are welcome to visit Dr Kung's website and send your thoughts and feedback to him.

e-mail   web@franciskung.net

http://www.franciskung.net

Book purchase and distribution details listed on the website.

www.ingramcontent.com/pod-product-compliance
Lightning Source LLC
Chambersburg PA
CBHW071403170526
45165CB00001B/164